Measuring Human Capabilities

AN AGENDA FOR BASIC RESEARCH ON THE ASSESSMENT
OF INDIVIDUAL AND GROUP PERFORMANCE POTENTIAL
FOR MILITARY ACCESSION

Committee on Measuring Human Capabilities:
Performance Potential of Individuals and Collectives

Board on Behavioral, Cognitive, and Sensory Sciences

Division of Behavioral and Social Sciences and Education

NATIONAL RESEARCH COUNCIL
OF THE NATIONAL ACADEMIES

THE NATIONAL ACADEMIES PRESS
Washington, D.C.
www.nap.edu

THE NATIONAL ACADEMIES PRESS 500 Fifth Street, NW Washington, DC 20001

NOTICE: The project that is the subject of this report was approved by the Governing Board of the National Research Council, whose members are drawn from the councils of the National Academy of Sciences, the National Academy of Engineering, and the Institute of Medicine. The members of the committee responsible for the report were chosen for their special competences and with regard for appropriate balance.

This study was supported by Contract/Grant No. W-911NF-12-1-0504 between the National Academy of Sciences and the Department of the Army. Any opinions, findings, conclusions, or recommendations expressed in this publication are those of the author(s) and do not necessarily reflect the views of the organizations or agencies that provided support for the project.

International Standard Book Number-13: 978-0-309-31717-7
International Standard Book Number-10: 0-309-31717-7
Library of Congress Control Number: 2015934162

Additional copies of this report are available from the National Academies Press, 500 Fifth Street, NW, Keck 360, Washington, DC 20001; (800) 624-6242 or (202) 334-3313; http://www.nap.edu.

Copyright 2015 by the National Academy of Sciences. All rights reserved.

Printed in the United States of America

Suggested citation: National Research Council. (2015). *Measuring Human Capabilities: An Agenda for Basic Research on the Assessment of Individual and Group Performance Potential for Military Accession.* Committee on Measuring Human Capabilities: Performance Potential of Individuals and Collectives, Board on Behavioral, Cognitive, and Sensory Sciences, Division of Behavioral and Social Sciences and Education. Washington, DC: The National Academies Press.

THE NATIONAL ACADEMIES
Advisers to the Nation on Science, Engineering, and Medicine

The **National Academy of Sciences** is a private, nonprofit, self-perpetuating society of distinguished scholars engaged in scientific and engineering research, dedicated to the furtherance of science and technology and to their use for the general welfare. Upon the authority of the charter granted to it by the Congress in 1863, the Academy has a mandate that requires it to advise the federal government on scientific and technical matters. Dr. Ralph J. Cicerone is president of the National Academy of Sciences.

The **National Academy of Engineering** was established in 1964, under the charter of the National Academy of Sciences, as a parallel organization of outstanding engineers. It is autonomous in its administration and in the selection of its members, sharing with the National Academy of Sciences the responsibility for advising the federal government. The National Academy of Engineering also sponsors engineering programs aimed at meeting national needs, encourages education and research, and recognizes the superior achievements of engineers. Dr. C. D. Mote, Jr., is president of the National Academy of Engineering.

The **Institute of Medicine** was established in 1970 by the National Academy of Sciences to secure the services of eminent members of appropriate professions in the examination of policy matters pertaining to the health of the public. The Institute acts under the responsibility given to the National Academy of Sciences by its congressional charter to be an adviser to the federal government and, upon its own initiative, to identify issues of medical care, research, and education. Dr. Harvey V. Fineberg is president of the Institute of Medicine.

The **National Research Council** was organized by the National Academy of Sciences in 1916 to associate the broad community of science and technology with the Academy's purposes of furthering knowledge and advising the federal government. Functioning in accordance with general policies determined by the Academy, the Council has become the principal operating agency of both the National Academy of Sciences and the National Academy of Engineering in providing services to the government, the public, and the scientific and engineering communities. The Council is administered jointly by both Academies and the Institute of Medicine. Dr. Ralph J. Cicerone and Dr. C. D. Mote, Jr., are chair and vice chair, respectively, of the National Research Council.

www.national-academies.org

COMMITTEE ON MEASURING HUMAN CAPABILIITES: PERFORMANCE POTENTIAL OF INDIVIDUALS AND COLLECTIVES

PAUL R. SACKETT (*Chair*), Department of Psychology, University of Minnesota, Minneapolis
GEORGIA T. CHAO, Eli Broad Graduate School of Management, Michigan State University
ANN DOUCETTE, The Evaluators' Institute, The George Washington University
RANDALL W. ENGLE, School of Psychology, Georgia Institute of Technology
RICHARD J. GENIK II, Department of Biomedical Engineering, Wayne State University College of Engineering, and School of Medicine, Detroit, MI
LEAETTA HOUGH, Dunnette Group, Ltd., Saint Paul, MN
PATRICK C. KYLLONEN, Center for Academic and Workforce Readiness and Success, Educational Testing Service, Princeton, NJ
JOHN J. MCARDLE, Department of Psychology, University of Southern California
FREDERICK L. OSWALD, Department of Psychology, Rice University
STEPHEN STARK, Department of Psychology, University of South Florida
WILLIAM J. STRICKLAND, Human Resources Research Organization, Alexandria, VA

CHERIE CHAUVIN, *Study Director*
TINA M. WINTERS, *Associate Program Officer*
RENÉE L. WILSON GAINES, *Senior Program Assistant*

BOARD ON BEHAVIORAL, COGNITIVE, AND SENSORY SCIENCES

SUSAN T. FISKE (*Chair*), Department of Psychology and Woodrow Wilson School of Public and International Affairs, Princeton University
LAURA L. CARSTENSEN, Department of Psychology, Stanford University
JENNIFER S. COLE, Department of Linguistics, University of Illinois at Urbana–Champaign
JUDY DUBNO, Department of Otolaryngology-Head and Neck Surgery, Medical University of South Carolina
ROBERT L. GOLDSTONE, Department of Psychological and Brain Sciences, Indiana University
DANIEL R. ILGEN, Department of Psychology, Michigan State University
NINA G. JABLONSKI, Department of Anthropology, Pennsylvania State University
JAMES S. JACKSON, Institute for Social Research, University of Michigan
NANCY G. KANWISHER, Department of Brain and Cognitive Sciences, Massachusetts Institute of Technology
JANICE KIECOLT-GLASER, Department of Psychology, Ohio State University College of Medicine
BILL C. MAURER, School of Social Sciences, University of California, Irvine
JOHN MONAHAN, School of Law, University of Virginia
STEVEN E. PETERSEN, Department of Neurology and Neurological Surgery, Washington University in St. Louis School of Medicine
DANA M. SMALL, Department of Psychiatry, Yale Medical School
TIMOTHY J. STRAUMAN, Department of Psychology and Neuroscience, Duke University
ALLAN R. WAGNER, Department of Psychology, Yale University
JEREMY M. WOLFE, Brigham and Women's Hospital, Departments of Ophthalmology and Radiology, Harvard Medical School

BARBARA A. WANCHISEN, *Director*
TENEE DAVENPORT, *Program Coordinator*

Preface

In the face of increasing pressure to improve the productivity of the Army's workforce, the U.S. Army Research Institute for the Behavioral and Social Sciences (ARI) approached the National Research Council (NRC) to develop an agenda for basic research on effective measurement of human capability with the goal of enhancing the military's selection and assignment process. ARI requested a research agenda to guide policy, procedures, and research related to the measurement of individual capability and the combination of individual capabilities to create collective capacity to perform.

In response to the request from ARI, the NRC established the Committee on Measuring Human Capabilities: Performance Potential of Individuals and Collectives, under the oversight of the Board on Behavioral, Cognitive, and Sensory Sciences. This report is the work of that committee and presents the committee's final conclusions and recommendations.

Members of the committee were volunteers carefully selected by the NRC to cover a spectrum of relevant academic specialties and to bring expertise in both basic research and practical applications. Several committee members have had significant experience with historical and current assessment programs utilized in the military enlistment process as well as outside the military.

The study was conducted in two phases over a 30-month period, during which the committee met a total of five times and hosted a public workshop, the summary of which was published in 2013. The study's first phase focused entirely on planning and hosting the public workshop, as well as the subsequent summary publication. The study's second phase was designed to allow the committee to consider specific research areas presented at the original workshop (as well as some areas that were not included due to time constraints or the availability of key presenters) in order to develop consensus findings and recommendations in accordance with the study's statement of task. In considering the most promising areas of research presented during the workshop, the committee membership was altered and expanded during the second phase to supplement the expertise of the original workshop planning committee members.

The recommendations presented in this report focus on an agenda for basic research that is likely to develop into a viable applied research program. In the course of preparing this report, each committee member took an active role in drafting chapters, leading discussions, and reading and commenting on successive drafts. The committee deliberated all aspects of this report, and its final content is the result of the members' tremendous effort, dedication, and interest in developing improved assessments of performance potential that are specifically relevant to the U.S. military services' selection and assignment processes.

Paul R. Sackett, *Chair*
Cherie Chauvin, *Study Director*
Committee on Measuring Human Capabilities:
Performance Potential of Individuals and Collectives

Acknowledgments

This study was sponsored by the U.S. Army Research Institute for the Behavioral and Social Sciences (ARI). The committee wishes to thank Gerald Goodwin, chief of foundational science at ARI, and his entire research team for their support and enthusiasm for this project. The committee also wishes to thank Paul Gade, George Washington University and formerly at ARI, for his continued interest in ARI's work and in this study in particular. The active participation of ARI representatives (current and former) in the committee's data gathering sessions played a crucial role in the committee's understanding of ARI's needs and interests. Specifically, these discussions aided the committee's awareness of the challenges and opportunities in the process of selection and assignment across the armed services.

In addition to the insightful presentations of invited experts during the study's public workshop held April 3-4, 2013, which are summarized in the National Research Council (NRC) report, *New Directions in Assessing Performance Potential of Individuals and Groups: Workshop Summary*, the committee benefited from several subsequent presentations that explored certain topics in more depth and detail. The committee thanks the following individuals for their expert presentations during the study's second phase: Christopher Codella, IBM; Mica Endsley, U.S. Air Force; Richard Landers, Old Dominion University; David Lubinski, Vanderbilt University; and Wim Van der Linden, CTB/McGraw-Hill. And finally, Dr. Michael I. Posner, Department of Psychology, University of Oregon (Emeritus), was appointed as an agent of the committee to provide subject matter expertise to assist the committee in verifying the scientific integrity of technical aspects of the final report.

Among the NRC staff, special thanks are due to Barbara A. Wanchisen, director, Board on Behavioral, Cognitive, and Sensory Sciences, who provided oversight to the study process. Additionally, special thanks to Tina Winters, associate program officer, who was instrumental in organizing data gathering opportunities for the committee and facilitating the development of the committee's final report. Renée Wilson Gaines, senior program assistant, also provided critical support to the study process and answered the committee's administrative and logistical needs. Ellen Kimmel and Rebecca Morgan, of the NRC Research Center, were an invaluable resource over the duration of the project as they identified relevant research to answer questions on historical precedent, current developments, and future forecasts. We also thank NRC consultant Robert Katt for final editing of the manuscript. And finally we thank the executive office reports staff of the Division of Behavioral and Social Sciences and Education, especially Kirsten Sampson Snyder, who managed the review process, and Yvonne Wise, who oversaw the final publication process.

This report has been reviewed in draft form by individuals chosen for their diverse perspectives and technical expertise, in accordance with procedures approved by the NRC's Report Review Committee. The purpose of this independent review is to provide candid and critical comments that will assist the institution in making its published report as sound as possible and to ensure that the summary meets institutional standards for objectivity, evidence, and responsiveness to the study charge. The review comments and draft manuscript remain confidential to protect the integrity of the deliberative process. We wish to thank the following individuals for their review of this summary: Janis Cannon-Bowers, Institute for Simulation and Training, University of Central Florida; Scott T. Grafton, Department of Psychological and Brain Sciences, University of California at Santa Barbara; Michael A. McDaniel, Human Resources and Organizational Behavior, Department of Psychology, Virginia Commonwealth University; James L. Mohler, College of Technology, Purdue University; James L. Outtz, President, Outtz and Associates, Washington, DC; Christopher Patrick, Psychology Department, Florida State University; Thomas Redick, Department of Psychological Sciences, Purdue University; Michael G. Rumsey, retired, Chief Personnel Assessment Research Unit, U.S. Army Research Institute for the Behavioral and Social Sciences; Ghebre (Gabe) E. Tzeghai, Corporate Research and Development and Innovation, Procter & Gamble Company; Howard M. Weiss, School of Psychology, Georgia Institute of Technology; and Carl E. Wieman, Department of Physics, Stanford University.

Although the reviewers listed above provided many constructive comments and suggestions, they were not asked to endorse the content of the report nor did they see the final draft of the report before its release. The review of this report was overseen by Daniel R. Ilgen, Department of Psychology and Management, Michigan State University, and Floyd E. Bloom, Molecular and Integrative Neuroscience Department, The Scripps Research Institute (Emeritus). Appointed by the NRC, they were responsible for making certain that an independent examination of this report was carried out in accordance with institutional procedures and that all review comments were carefully considered. Responsibility for the final content of this report rests entirely with the authoring committee and the institution.

Paul R. Sackett, *Chair*
Cherie Chauvin, *Study Director*
Committee on Measuring Human Capabilities:
Performance Potential of Individuals and Collectives

Contents

Executive Summary — 1

SECTION 1: INTRODUCTION

1 Overview — 7
Setting the Stage: The Current Army Enlisted Soldier Accession System, 8
Research Approach, 10
An Overview of the Research Agenda, 19
References, 24

SECTION 2: IDENTIFICATION AND MEASUREMENT OF NEW PREDICTOR CONSTRUCTS

2 Fluid Intelligence, Working Memory Capacity, Executive Attention, and Inhibitory Control — 29
Fluid Intelligence, 30
Working Memory Capacity and Executive Attention, 32
Inhibitory Control, 40
References, 43

3 Cognitive Biases — 53
Examples of Cognitive Biases in Action, 54
The Nature and Diversity of Cognitive Biases, 56

IARPA's Cognitive Bias Mitigation Program, 59
Future Research, 60
Research Recommendation, 62
References, 63

4 **Spatial Abilities**   65
Defining Spatial Abilities, 66
Testing Spatial Abilities for Military Entrance, 67
State of the Science, 70
A Path Forward, 79
Research Recommendation, 80
References, 80

SECTION 3: IDENTIFICATION AND PREDICTION OF NEW OUTCOMES

5 **Teamwork Behavior**   87
Team Outcomes: Defining Team Effectiveness, 89
Teamwork Processes and Emergent States of Teams, 91
Team Member Inputs: Selecting and Classifying Individuals for Effective Teams, 94
A Path Forward, 98
Research Recommendation, 100
References, 100

SECTION 4: HYBRID TOPICS WITH JOINT FOCUS ON NEW PREDICTOR CONSTRUCTS AND PREDICTION OF NEW OUTCOMES

6 **Hot Cognition: Defensive Reactivity, Emotional Regulation, and Performance under Stress**   107
Defensive Reactivity/Fearfulness versus Fearlessness/Boldness, 109
Emotion Regulation, 112
Performance under Stress, 118
Research Recommendation, 121
References, 122

7 **Adaptability and Inventiveness**   127
Background, Definitions, and Issues, 127
Adaptability/Inventiveness as an Outcome Variable, 131

Adaptability/Inventiveness as an Individual-Difference
    Cognitive Variable: Idea Production Measures Increment Validity
    Over General Cognitive Ability, 132
Adaptability/Inventiveness as an Individual-Difference
    Noncognitive Variable, 133
Research Recommendation, 152
References, 153

SECTION 5: METHODS AND METHODOLOGY

8 Psychometrics and Technology   161
   Introduction, 161
   Historical Background, 162
   Future Research Investments, 167
   Research Recommendation, 177
   References, 178

9 Situations and Situational Judgment Tests   187
   Validity, 188
   Instructions and Format, 190
   Subgroup Differences and Adverse Impact, 193
   Psychometric Findings, 194
   Future Directions, 197
   Research Recommendation, 199
   References, 199

10 Assessment of Individual Differences Through
   Neuroscience Measures   203
   Introduction, 203
   Monitoring Test Performance, 205
   Designing Better Tests and Testing Environments, 207
   Predicting Human Performance, 208
   Research Recommendation, 210
   References, 211

SECTION 6: THE RESEARCH AGENDA

11 The Research Agenda   217
   Implementation, 217
   Research Topics: Committee Conclusions and Recommendations, 219

APPENDIXES

A	Workshop Agenda and Selection of Additional Topics Considered for Workshop Agenda	233
B	Phase II Data Gathering Presentations	239
C	Biomarkers	241
D	Neural Signals and Measurement Technologies	247
E	Biographical Sketches of Committee Members and Staff	259

Executive Summary

Each year, the U.S. Army must select from an applicant pool in the hundreds of thousands to meet annual enlistment targets, currently numbering in the tens of thousands of new soldiers each year. A critical component of the selection process for enlisted service members is the formal assessments administered to applicants to determine their performance potential. All applicants passing basic educational attainment and moral character screens take the Armed Services Vocational Aptitude Battery (ASVAB), a cognitive knowledge, skill, and ability battery of ten tests used by all branches of the armed services. A subset of four of these tests focusing on verbal and mathematical skills is used to create a composite known as the Armed Forces Qualifying Test (AFQT), which is used as a basic entry screen. In addition, Army applicants scoring below the 50th percentile on the AFQT also take the Tailored Adaptive Personality Assessment System (TAPAS), a personality assessment used to predict job performance and risk of attrition. The results from these assessments determine whether or not a candidate may proceed in the application process; they are also used to inform decisions about assignment of new enlisted soldiers to occupational specialties.

While the validity evidence supporting both the ASVAB and TAPAS is very strong, the ability and personality domains they measure are not the sole determinants of key outcomes such as job performance and attrition. Thus, the ASVAB and TAPAS do not predict these outcomes perfectly, raising question as to whether prediction could be improved further by supplementing the ASVAB and TAPAS results with measures of additional individual-differences constructs. Given the large numbers of individuals screened each year and the high costs of decision errors in screening potential soldiers (e.g., the high costs of recruiting and training a soldier who subsequently performs poorly or leaves the service prior to completing a tour of duty), even small increases in the predictive accuracy of a selection system can be of great value.

In light of these considerations, the U.S. Army Research Institute for the Behavioral and Social Sciences approached the National Research Council to develop an agenda for basic research to maximize the efficiency, accuracy, and effective use of human capability measurement in the military's selection and initial occupational assignment process. In response to this request, the National Research Council established the Committee on Measuring Human Capabilities: Performance Potential of Individuals and Collectives.

The committee operated under a number of constraints, focusing on (1) attributes broadly useful for first-term enlisted soldiers and for which there is a theoretical foundation for the measurement of the attribute and for its relevance to important military outcomes; and (2) measures that can be administered pre-accession in a cost-effective manner to large numbers of candidates without requiring elaborate equipment or special skills. Within these constraints, and after a careful review of the research literature and consideration of presented material, the committee arrived at a basic agenda for research presented in four sections as follows.

IDENTIFICATION AND MEASUREMENT OF NEW PREDICTOR CONSTRUCTS

Fluid intelligence, working memory capacity, executive attention, and inhibitory control. Fluid intelligence reflects the ability to think logically and develop solutions when faced with novel or unfamiliar problems. Working memory capacity is the cognitive function that enables individuals to hold information in mind and simultaneously manipulate that information or other information. The related function of executive attention is the ability to prevent attention capture by other endogenous and exogenous events. Similarly, inhibitory control involves the ability to resist distractions and control one's responses.

Cognitive biases. Cognitive biases refer to ways of reflexive thinking that can produce errors in judgment or decision making or produce departures from the use of normative rules or standards.

Spatial abilities. Spatial abilities involve the capacity to understand an object's (including one's own) spatial relationship to and within surroundings and to understand representations of multidimensional figures in one-dimensional displays.

IDENTIFICATION AND PREDICTION OF NEW OUTCOMES

Teamwork behavior. This topic covers individual attributes and team factors that may be of use in predicting team success.

HYBRID TOPICS WITH JOINT FOCUS ON NEW PREDICTOR CONSTRUCTS AND PREDICTION OF NEW OUTCOMES

Hot cognition: defensive reactivity, emotional regulation, and performance under stress. Hot cognition refers to how individuals perform in situations that elicit strong emotions (in contrast to cognition under circumstances of cool, level, or moderate emotions, or "cold cognition").

Adaptability and inventiveness. Adaptability involves the ability to adjust and accommodate to changing and unpredictable physical, interpersonal, cultural, and task environments. Inventiveness refers to the ability to think innovatively and produce novel high-quality and task-appropriate ideas, incorporating an orientation toward problem solving.

METHODS AND METHODOLOGY

Psychometrics and technology. Psychological measurement methodology and data analytics, especially those made possible through advances in technology, offer new opportunities and challenges with potential for high payoffs in testing efficiency and effectiveness.

Situations and situational judgment tests. Situational judgment tests are measures that assess individuals' abilities to use judgment to interpret, evaluate, and weigh alternate courses of action appropriately and effectively.

Assessment of individual differences through neuroscience measures. Within the vast field of neuroscience, science-based strategies for monitoring neural activity may be useful for yielding important information about factors underlying candidates' test performance or testing the validity of various assessment strategies.

Section 1

Introduction

1

Overview

This report represents the second phase of a larger project on measuring human capabilities in the context of developing a basic research agenda aimed at identifying possible predictors of first-term soldier performance in the U.S. Army that could usefully supplement measures currently used as screening devices in the enlistment process. In Phase 1, committee members identified promising research topics, identified experts in these topic areas, and convened a workshop at which invited experts presented their research. That workshop is summarized in a report entitled *New Directions in Assessing Performance Potential of Individuals and Groups: Workshop Summary* (National Research Council, 2013). In Phase 2, the committee was enhanced with additional members and charged with further exploration, building on the workshop, in order to develop a recommended future research program for the Foundational Science Research Unit of the U.S. Army Research Institute for the Behavioral and Social Sciences (ARI). Box 1-1 provides the specifics of the committee charge.

The committee interpreted this charge as requesting recommendations for a basic research agenda aimed at identifying ways to supplement the Army's enlisted soldier accession system with additional predictors that go beyond the currently used cognitive and personality measures (described in the following section) and that have the potential to improve the already high quality of accession decisions. To accomplish this task, the committee has focused on potential determinants of individual and collective (e.g., team) performance.

> **BOX 1-1**
> **Charge to the Committee**
>
> The charge to the committee from ARI was as follows:
>
> In Phase 2, the committee will consider in more depth and detail specific research areas presented at the original workshop. The committee will develop consensus findings and recommendations to guide policy, procedures, and research related to the measurement of individual capability and the combination of individual capabilities to create collective capacity to perform. To the extent the evidence warrants, the committee will identify the most promising research areas to assess through the following questions and tasks:
>
> 1. What are the most promising approaches to measurement of individual capability and the combination of individual capabilities to create collective capacity to perform? Do recent or emerging theoretical, technological, and/or statistical advances provide scientifically valid new approaches and/or measurement capabilities?
> 2. Assess the neuroscience advances related specifically to the understanding of individual differences that suggest new ways to approach empirical research and theory development in this area. How should the U.S. Army Research Institute (ARI) take advantage of these in its basic research program?
> 3. Recommend a future research agenda for ARI to maximize the efficiency, accuracy, and effective use of human capability measurement related to theories of individual differences (cognitive, affective, personality, social or interpersonal skills), testing and measurement methods, test theory, statistical and mathematical modeling of collective/group/team performance, and the combination of individual capabilities to create collective capacity to perform. In developing this research agenda, the committee will identify immediate research opportunities in the most promising topics likely to have the highest near-term payoff.
> 4. Specify the basic research funding level needed to implement the recommended agenda for future ARI research.

SETTING THE STAGE:
THE CURRENT ARMY ENLISTED SOLDIER ACCESSION SYSTEM

The committee's focus is on the initial selection process that determines eligibility for entry-level positions within the Army. Individuals with an interest in enlisting must meet standards in a number of areas. These include two formal testing vehicles: The Armed Services Vocational Aptitude Battery (ASVAB) is a cognitive knowledge, skill, and ability battery of tests. The Tailored Adaptive Personality Assessment System (TAPAS) is a computer-administered personality measure. In addition, there are three nontest screens: educational attainment, an examination of moral character, and an examination of physical and medical readiness to serve. A concise review of the U.S. military's selection and assessment system is provided by Sellman and colleagues (2010).

The ASVAB is primarily conducted as the CAT-ASVAB, a computer-administered adaptive version of the ASVAB test battery comprising 10 tests. Four of the tests (word knowledge, paragraph comprehension, arithmetic reasoning, and mathematics knowledge) are combined into a composite known as the Armed Forces Qualifying Test (AFQT), which is used as a basic entry screen. The other tests (general science, electronics information, mechanical comprehension, auto information, shop information, and assembling objects) are used for determining qualification for specific occupational assignments once the entry screen has been passed. Extensive research supports the predictive capability of ASVAB performance for subsequent training and first-term job performance (see Armor and Sackett, 2004, for a review).

The TAPAS is a personality measurement system that can be configured to deliver nonadaptive and adaptive personality tests based on Item Response Theory (IRT). The TAPAS tests typically measure 12-18 of a possible 28 narrow personality factors. TAPAS factor scores are used to form composites, currently called "can do," "will do," and "persistence," which are used to predict job performance and attrition criteria. While the ASVAB is part of the enlistment decision for all applicants, the TAPAS is administered to those scoring below the 50th percentile on the AFQT. Those below the 50th percentile have a higher risk of failure to meet standards and adjust successfully to military life, and the TAPAS is used to identify and screen out high-risk candidates.

The Army also classifies applicants on the basis of educational attainment, as extensive research shows that possession of a high school diploma is the best single predictor of successful adjustment to military life (Strickland, 2005; Trent and Laurence, 1993). The rate of noncompletion of a tour of duty is markedly higher for high school dropouts and for holders of other credentials such as a General Education Development certificate or high school completion via home schooling than for diploma holders (Strickland, 2005). Thus, applicants are put into one of three tiers (tier 1: diploma; tier 2: alternate credential; tier 3: dropout), with more stringent AFQT standards applied below the first tier.

Applicants with a qualifying AFQT score receive a physical examination covering a range of features including blood pressure, pulse, visual acuity, hearing, blood testing, urinalysis, and drug and HIV testing. Some conditions require medical treatment before enlistment; others are disqualifying, though applicants can apply for a waiver in some circumstances.

Finally, applicants must meet moral character standards. Some criminal activities are immediately disqualifying. In other cases, applicants can apply for a waiver, which prompts a review of the specific circumstances and a case-specific determination as to whether the applicant will be permitted to enlist.

RESEARCH APPROACH

Constraints

The committee operated under a number of constraints as it examined possible additional screening tools that could usefully augment the current testing in the cognitive ability and personality domains. These constraints were determined via instructions from and discussions with representatives of the Foundational Science Research Unit, ARI, and the committee has taken them as conjoint conditions (acting together) on what should be included in the focus of our report and what should be minimized or excluded altogether.

The first constraint was that the committee should focus on attributes broadly useful for first-term enlisted soldiers. We thus excluded attributes that are relevant to just a single occupational specialty or to a select set of occupational specialties. This constraint is consistent with the cognitive and personality measures currently used by the Army: cognitive problem solving skill and a pattern of personality attributes reflecting ability to adjust to military life are broadly relevant regardless of occupational specialty.

Second, the committee was instructed to focus on measures that can be administered pre-accession and in a cost-effective manner to large numbers of individual candidates without requiring special skills to administer the measure or to evaluate performance on the measure, and without requiring elaborate equipment. These constraints preclude consideration of predictor measures such as:

- Measures involving complex work sample measures (with extensive sampling of skill performance and work competencies possibly requiring days or weeks to complete);
- Measures involving interpersonal interaction (e.g., role plays, team performance tasks);
- Neurological measures (including invasive procedures);
- Physical ability and fitness measures; and
- Assessments that can play a useful role for the Army but that do not take place pre-accession, such as mid-career assessments or post-injury return-to-work assessments.

The neurological measures category listed above merits amplification. While the constraint regarding a restriction to measures that do not require special skill and/or equipment to administer precludes neurological measures for present use as a routine part of entry level screening, we do, consistent with the charge to the committee, consider potential roles for neurological measures (see Chapter 10). First, we consider them for use as criterion measures against which other measures (e.g., self-reports) can be evaluated. Second, we consider them for use in settings other than mass screening of candidates. For example, there may be roles for neurological measures in follow-up assessment of limited subsets of candidates, such as those producing a particular and difficult to interpret pattern of results on other measures used in the screening process. Finally, we consider the possibility for measurement developments that may in the future make large scale screening with one or more neurological measures logistically possible.

The third constraint was to focus on attributes for which there is a theoretical foundation for the measurement of the attribute of interest and for the relevance of the attribute to important military outcomes. This constraint precluded consideration of approaches based on brute empiricism, such as the use of empirical biographical data keys in which various background characteristics are assigned weights based on the degree to which they prove to differentiate between soldiers who score high versus low on outcomes of interest. Similarly, consideration of features such as birth order were excluded because a strong theoretical foundation for their use is lacking.

Finally, we received specific instruction that genetic screening was outside the committee's purview. Thus, we did not include genetic testing in our review.

Note that, within these constraints, measures excluded from consideration as possible supplements to the accession system may still be recommended for use as criterion measures in the evaluation for operational utility of other measures that are consistent with the constraints. For example, measures involving team performance tasks, excluded as a testing measure under the second constraint above, may nonetheless be useful as criteria against which individual measures of propensity for effectiveness as a team member may be evaluated. Furthermore, as the committee developed an agenda for future research, some topics were considered based upon the committee's expectations of the impact of future technology or other capabilities that could significantly change the feasibility for operational use in the long term.

Terminology

To understand potential improvements in human capability measurement, several important terms need to be understood, as they define what and how measurements are conducted. The committee uses the term "construct" to refer to the attribute label attached to a measure (e.g., arithmetic reasoning, fluid intelligence, conscientiousness). Other terms used in various places to refer to individual-differences measures include "trait" and "factor." "Trait" connotes a reasonably stable attribute (as opposed to a "state" such as mood or emotion, which is expected to change frequently). "Factor" denotes an attribute in common among a set of trait measures, identified through application of the technique of factor analysis.

Identifying Topics for Research

At early meetings during Phase 1 of the project, the committee identified a lengthy list of possible topics for research. Committee discussion led to identifying a subset of this list as worthy of further investigation. A sizable number of the selected topics were the focus of a workshop held in April 2013. Some selected topics could not be covered in the workshop due to time constraints or unavailability of the targeted speakers. Additional speakers were invited to subsequent committee meetings to address such topics. As Phase 2 of the project involved an expanded committee, we revisited the initial topic list from Phase 1, amending it to include input from new committee members. Committee membership included individuals with broad expertise in personnel selection in both civilian and military contexts, individual differences, performance measurement, teamwork, psychometrics, and neuroscience. Initial topic identification relied on the expertise and judgment of committee members. We asked questions such as "what is in operational use in other employment settings?", "what looks promising in the selection literature on new predictor constructs and/or new predictor methods?", and "what looks promising in the individual-differences literature that might prove applicable to personnel selection settings?"

We had available to us useful summaries of work addressing these questions. For example, the *Annual Review of Psychology* commissions systematic and thorough reviews of developments in the area of personnel selection on a recurring basis. The two most recent reviews at the time of the committee's work were by Hough and Oswald (2000) and by Sackett and Lievens (2008). Hough, Oswald, and Sackett serve on the present committee.

The committee used workshop content, subsequent presentations to the committee, review of publically available research and data, and discussions within the committee to identify the set of topics discussed in this report.

(For more details of the topics considered in developing this report, see the workshop agenda and selection of topics considered for the workshop agenda in Appendix A and the list of the data gathering presentations delivered to the committee during the study's second phase in Appendix B.) We do not offer an assessment of the topics that were considered but not included in the recommended research agenda. We acknowledge that the set of topics selected represent the collective judgment of the committee. It is possible that a differently constituted committee would identify some additional topics or would choose not to focus on some of the topics covered here. The committee membership does reflect broad and varied expertise relevant to our charge, and we are confident that we have identified a promising, even if not exhaustive, research agenda.

In evaluating research topics, we applied the following decision process. First, could we identify a conceptual basis for a linkage between a particular predictor construct and a criterion construct that can be expected to be of interest to the Army? Success of ARI's basic research program is largely determined by the feasibility of developing foundational science into applied research programs and ultimate implementation to affect U.S. Army policy and procedure. If the committee could not identify potential utility in the basic research results to improve prediction of soldier success, the topic was not considered further. Note that we did not view our task as limited to existing operational Army criteria (e.g., criteria used to assess training performance or attrition). A conceptually meaningful criterion construct, such as team effectiveness, could be considered even if a measure based on that construct is not currently in operational use.

Second, could we identify settings where we could see analogs to military performance, such as job performance in the civilian workforce, where measures of particular predictor constructs have been (a) successfully developed, (b) shown to be linked to criteria of interest, and (c) shown to have incremental validity over measures in the ability and personality domain? Although topics were not discarded from further consideration solely on the basis of failure to meet all three conditions, the committee weighed topics against each other, and topics included in this final report were judged to meet an appropriate minimum threshold given the prior research and data available on the particular topic.

Third, we sought to identify research developments that suggest a reconsideration of long-standing research domains that may have been rejected in the past for a variety of possible reasons. In particular, the committee sought constructs with proven predictive capability but that were not conducive to testing through standard paper and pencil tests. For example, we considered whether there are new measurement developments that could potentially overcome obstacles to the measurement of a particular predictor construct (e.g., the development of measurement methods more resistant to faking and coaching). The committee also evaluated research domains that may have been stymied due to lack of funding or due to misunderstood research results that may have deterred further research programs, as well as domains that may have been considered high-risk (and potentially high-payoff) compared to other research domains.

Fourth, we considered whether there are constructs for which a promising research base is developing but which have not been investigated in the context of personnel selection. This involved considering the broader individual-differences literature, rather than focusing solely on the personnel selection literature.

Fifth, one key feature that might easily be overlooked is that the charge to the committee focused on identifying a basic research agenda that might in time lead to improvements in the Army enlisted soldier selection process. Thus our charge excluded possible methods of improving selection that were, in the committee's judgment, beyond the basic research stage. Perhaps the most vivid example of this is the domain of vocational interest measurement. Vocational interest measures have for some time been viewed as useful to individuals for career guidance but of limited value for personnel selection. There has been a recent resurgence of research on the relationship between vocational interests and subsequent performance outcomes (e.g., Nye et al., 2012; Van Iddekinge et al., 2011), suggesting stronger interest-outcome relationships than had been seen in the past. The committee gave careful attention to this domain, including an invited presentation to the committee on the topic. After extensive discussion, however, the committee concluded that what was needed was a program of criterion-related validation research to determine whether this positive pattern of relationships would also be found in Army settings. Such work is essentially operational, as well-developed measures exist ready for tryout. Thus, while the committee is cautiously optimistic that vocational interest measurement has the potential to improve selection, the consensus was that this was not a basic research issue.

As the committee deliberated the list of possible topics, these questions were carefully considered to determine whether a possible topic satisfied a minimum threshold for inclusion in the final recommended research agenda. They also contributed to the decision process whereby topics were evaluated against each other so as to select the strongest candidates, by the committee's judgment, to be most likely to have the largest impact on improving the military personnel testing, selection, and assignment process. No single research topic was a perfect fit to all the criteria. Furthermore, large variations in prior research volume, strategy, and results were found between topics, and this is reflected in the presentation of those topics in the individual chapters of this report.

A Taxonomic Structure for Ways to Improve Selection Systems

Sackett and Lievens (2008) offered a taxonomy of ways that a selection system can be improved, and Sackett presented a version of this taxonomy at the workshop convened as part of Phase 1 of the current project. In particular, Sackett and Lievens (2008) proposed that a selection system can be improved by one or more of the following:

a. Identification and measurement of new predictor constructs;
b. Identification and prediction of new outcomes;
c. Improved measurement of existing predictor constructs; or
d. Identification of features that moderate predictor-criterion relationships (e.g., identifying circumstances under which predictor-criterion relationships are stronger or weaker).

The committee used this taxonomic system in considering potential research investments. While Sackett and Lievens used the terms "new constructs" and "existing constructs" in the context of the entire field of personnel selection, we view them in terms of constructs currently in use for Army enlisted soldier selection. For example, while spatial ability is included in the ASVAB, it is not currently in use for enlisted selection, and thus we view spatial ability as a new construct for consideration. Our recommendations fall into all four of the categories in the above taxonomy, and we structure the report in terms of these categories.

What emerged as the most prominent of the categories in this taxonomy is the identification and measurement of new predictor constructs. Thus, following this introductory chapter, Section 2 of the report contains chapters that describe fluid intelligence, working memory capacity, executive attention, inhibitory control, cognitive biases, and spatial abilities. Each of these domains is described in more detail below.

Another prominent category is the identification and prediction of new outcomes. Although we identify three new performance domains that are conceptually relevant for a broad range of Army enlisted soldier positions, Section 3 presents only the first of these: teamwork behavior. Note that investigations into the prediction of new outcomes may result either in a determination that these outcomes are well predicted by currently used predictor measures or in a determination that a new predictor or predictors are needed to predict these outcomes. Two chapters contain elements that cut across aspects of the previous two sections, and therefore Section 4 contains hybrid topics with joint focus on new predictor constructs and prediction of new outcomes. The first chapter in that section, hot cognition, describes the two constructs, defensive reactivity and emotion regulation, and one outcome, performance under stress. The second chapter in Section 4 also presents two closely linked topics— adaptability and inventiveness—which can be conceptualized either as an outcome variable to be predicted or as a predictor construct.

Section 5 contains single chapters linked to other domains in the taxonomy. A chapter on psychometrics focuses on both ways of measuring existing constructs better (e.g., using new developments in IRT to further improve the ASVAB) and on potential new measurement methods (e.g., gaming). A chapter on situational judgment discusses a measurement method that can potentially be used for improved measurement of existing constructs (e.g., measuring personality constructs) and measuring new constructs not currently part of the Army's enlisted soldier selection system. Finally, a chapter on neuroscience focuses broadly on the potential use of neuroscience-based measures as markers of psychological states (e.g., undue anxiety while completing existing Army selection instruments). These states may moderate predictor-criterion relationships, as candidates exhibiting undue anxiety may produce test scores that are systematically lower than their true standing on the construct of interest.

The report concludes with a single chapter in Section 6, The Research Agenda. For the convenience of the reader, Chapter 11 includes a consolidated list of the committee's conclusions and recommendations, which together comprise the recommended research agenda for ARI's Foundational Science Research Unit. This final chapter also presents the committee's assessment of the funding level needed to implement the recommended research agenda.

Considerations in Choosing Criteria

A widely accepted principle within the field of personnel selection is that to develop a selection system, one must begin by specifying the criterion of interest. Using a simple example, if told "we want a selection system for supermarket cashiers," the response is to question the organization further: do you want cashiers who are fast in scanning groceries, friendly in dealing with customers, or reliable in their attendance? Some firms may emphasize speed and efficiency; others may emphasize friendliness. Some may want a balance between speed, efficiency, and friendliness. This has implications for the subsequent selection system: the individual attributes that predict who will be quick in scanning groceries are likely to be very different from those that predict warm and friendly customer interactions.

Importantly, the choice to, for instance, focus on predicting speed and efficiency versus friendly customer interaction is a matter of organizational values. It is not appropriate for the selection researcher to assert that the organization should value one outcome versus the other. The researcher can inform the organization about the degree to which a given outcome is predictable, but the choice of the outcome(s) of interest is ultimately a matter of organizational strategy.

These ideas have major implications for the recommendations developed in this report. The charge to the committee was to identify a research agenda with the potential of improving the Army's enlisted soldier selection system. This is a very broad charge. The committee would have acted very differently had it been presented with a charge that focused on a single specific criterion: for example, improve soldier's technical proficiency or reduce the rate of discharge for disciplinary reasons or reduce the rate of attrition due to lack of adjustment to military life. We also would have acted differently had our charge been to focus on selection criteria for classification of individuals into occupational specialties or specific jobs. However, absent this advance specification, we considered prospects for improving the selection system regarding a wide range of criteria.

That there is interest in multiple criteria in military selection is reflected in the currently used selection tools. At a high level of abstraction, the job performance domain can be subdivided into "can do" and "will do" domains. The ASVAB focuses on the "can do" domain: it is an effective predictor of the degree to which an enlistee will become technically proficient following training. It is not a particularly effective predictor of the typical degree of effort an enlistee will exert, or of the degree to which an enlistee will avoid behaviors that would result in disciplinary action. In contrast, the personality domains measured by the TAPAS includes a focus on the "will do" domain, and the TAPAS is predictive of avoiding disciplinary action and effective adjustment to military life. (For a recent discussion of the broad array of individual-differences constructs relevant to the military, see Rumsey and Arabian, 2014.)

Thus, the Army has interest in multiple criteria. Army research on the use of individual-differences measures that predict outcomes of interest has examined a wide range of criteria, including task proficiency, effort, maintaining military discipline, adjustment to military life, and attrition, among others. Therefore, the committee cast a broad net in developing recommendations for research. The requirement that we set for ourselves was that we could see a conceptual or empirical link between an attribute under consideration and one or more outcomes that constitute a component of overall individual or team effectiveness.

In considering outcomes of interest, we were informed by ongoing conceptual and empirical work about the underlying structure of individual and team effectiveness. A variety of scholars have advocated for differing representation of the underlying dimensionality of individual and team effectiveness. Campbell (2012) summarizes and integrates a variety of perspectives in the structure of behavior, performance, and effectiveness in contemporary organizations. We drew from a number of these perspectives, rather than embracing a single approach. We outline here a set of outcome variables that we believe are broadly relevant for organizations in general and the Army in particular.

Task proficiency. This is the degree to which individuals perform substantive tasks that are part of one's job. Many tasks may be specific to that job, but there are also likely to be common tasks that cut across jobs.

Demonstrating effort. This involves consistency of effort, willingness to put in extra time and effort when required, and willingness to persist under adverse conditions.

Maintaining personal discipline. This involves the avoidance of negative and counterproductive behavior, such as rule infraction and illegal behavior.

Facilitating peer and team performance. This involves supporting, helping, and informally training peer team members; serving as a role model; and helping keep the team directed and on task. These are components of what is commonly termed "citizenship" in the organizational literature.

Adaptive performance. This involves multiple subfacets, including handling stressful emergency or crisis situations; facing uncertain situations and solving problems creatively; and dealing effectively with changes in organizational goals, individual performance requirements, and the work environment.[1]

Adjustment to military life. This involves dealing effectively with the transition from civilian life to the military environment (e.g., a structured, hierarchical setting; restriction on personal choice; living in close quarters with others; and physical demands; among others).

Attrition. This can reflect voluntary or involuntary departure from the Army prior to completion of a contracted tour of duty. While often used as a criterion measure, it can be viewed as reflecting one or more

---

[1] Recently, Army leaders, such as Lt. Gen. Robert Brown, commander of the Army Combined Arms Center, have referred to the need for soldiers who "improve and thrive in conditions of chaos" (see *Army Times* article on "The Human Dimension" panel during the 2014 Association of the United States Army convention, available at http://www.armytimes.com/article/20141015/NEWS/310150065/Wanted-Soldiers-who-thrive-chaos [October 2014]).

of the more specific outcome variables above (e.g., voluntary turnover as a result of failure to adjust to military life, involuntary turnover as a result of serious rule infraction).

In Table 1-1, we present a grid that pairs each of the research domains for which we offer recommendations with this set of outcomes. For each research domain, we identify the outcome or outcomes for which we view a linkage as plausible. We do not view this as etched in stone; arguments that a domain may be linked to additional outcomes are possible. One reason for providing this grid is to show that each domain is linked to one or more outcomes, which is the basis for that domain being included as part of our proposed research agenda.

There is a second critical implication of this grid. Some may ask why we do not prioritize our recommendations (e.g., rank them 1-10). The reason is linked to the point developed earlier in this section that the choice of the outcome measure(s) on which to focus is a matter of organizational values, rather than a scientific question. Should the Army decide that any one of the outcomes in the grid is strategically of greatest value to its mission(s), then research domains linked to those outcomes would become higher in priority. Furthermore, particular occupational specialties might place greater value on different outcomes, thereby giving certain research domains priority for both selection and classification purposes. Put another way, the research domains we identify could be prioritized very differently depending on the value that the Army assigns to each outcome domain.

AN OVERVIEW OF THE RESEARCH AGENDA

In Chapters 2 through 10, divided into four sections, the committee presents a summary of available research and the committee's assessment of that research in consideration of a future research agenda to improve selection and retention of successful soldiers. The research domains presented in each chapter are outlined below.

Section 2. Identification and Measurement of New Predictor Constructs

The report's second section includes three chapters that present future research opportunities in the identification and measurement of new predictor constructs.

TABLE 1-1 Grid Showing Links Between Research Domains and Outcomes

	Task Proficiency	Demonstrating Effort	Personal Discipline	Peer and Team Performance	Adaptive Performance	Adjustment to Military Life	Attrition
Fluid Intelligence, Working Memory Capacity, Executive Attention, and Inhibitory Control	x				x		x
Cognitive Biases	x				x		x
Spatial Abilities	x						x
Teamwork Behavior		x		x	x		x
Hot Cognition and Performance Under Stress	x	x	x		x	x	
Adaptability and Inventiveness					x		x
Psychometrics and Technology	x	x	x	x	x	x	x
Situational Judgment Tests	x	x	x	x	x	x	x
Neuroscience	x	x	x	x	x	x	x

Chapter 2. Fluid Intelligence, Working Memory Capacity, Executive Attention, and Inhibitory Control

Chapter 2 discusses fluid intelligence, working memory capacity, executive attention, and inhibitory control in relation to an individual's emotional, behavioral, and impulse control. Many intelligence measures focus on crystallized intelligence: the learned and acquired skills and knowledge component of intelligence. Assessment of fluid intelligence could potentially reveal more about an individual's reasoning and novel problem-solving abilities. Working memory capacity and executive attention assessments, which are relatively short and easy to administer, have been found to be valid in predicting performance on a large variety of real-world cognitive tasks.

Chapter 3. Cognitive Biases

Chapter 3 describes cognitive biases that can produce errors in judgment or decision making. For example, projection (assuming others share one's own feelings, attitudes, and values) can interfere with soldiers' abilities to accurately judge the motives of others, such as host-nation citizens or international coalition military members. Cognitive biases operate in both everyday reasoning and decision making and also may play a role in life-and-death disasters; therefore, learning about individuals' susceptibility or proneness to cognitive biases may be useful for informing assignment decisions. One important question in this area is the degree to which cognitive biases can be mitigated by training.

Chapter 4. Spatial Abilities

Chapter 4 considers the spectrum of skills in the domain of spatial abilities. Research suggests that spatial abilities may be an important predictor of performance, particularly in scientific and technical fields. Multiple facets of spatial abilities have been identified or proposed, all of which relate to the many different ways individuals understand their own spatial relationship to and within surroundings and also the way individuals understand representations of multidimensional figures in one-dimensional displays. Although one spatial ability measure (Assembling Objects) is included in the ASVAB, this chapter presents evidence of the potential value of other approaches to the measurement of spatial abilities that may yield more useful information for military selection and classification.

Section 3. Identification and Prediction of New Outcomes

The third section consists of a single chapter on Teamwork Behavior, one of three new outcomes the committee identified with potential for identification and prediction in military assessment settings. The other two outcomes are discussed as part of hybrid chapters in Section 4.

Chapter 5. Teamwork Behavior

Chapter 5 considers individual and team factors that may be of use in predicting successful teamwork behavior. The chapter focuses on how selection and classification of entry-level enlisted soldiers can improve unit performance and mission success. The Input-Process-Outcome model can serve as a loose framework to identify future research objectives. Starting with the end goal, the committee first discusses team outcomes to define the criteria domain for selection and classification. Next, we examine team processes and emergent states as more proximal criteria of collective capacity. Finally, we examine how future research on individual-level inputs to teams might help understand who is best suited for teamwork and how individuals might be better classified into specific Army small units, to include teams, squads, and platoons.

Section 4. Hybrid Topics with Joint Focus on New Predictor Constructs and Prediction of New Outcomes

The report's fourth section includes two hybrid topics with aspects that cut across the two previous sections and thereby represent both new predictor constructs and the prediction of new outcomes.

Chapter 6. Hot Cognition: Defensive Reactivity, Emotional Regulation, and Performance Under Stress

Chapter 6 examines "hot cognition": how individuals perform in situations that elicit strong emotions (in contrast to cognition under circumstances of cool or moderate emotions, or "cold cognition"). Hot cognition is responsible for such behaviors as defensive reactivity: the degree to which one is prone to negative emotional activity (particularly fear) in threatening situations. Fear is often an unproductive emotion, especially for combat soldiers, whereas fearlessness or boldness can be a productive emotion. However, when taken too far, fearlessness might be maladaptive, contributing to a soldier's disregard for safety procedures or operational protocol. Research opportunities in this domain include, for example, investigating whether there may be an optimal level of defensive reactivity for performance in particular conditions or by a particular individual.

Chapter 7. Adaptability and Inventiveness

Chapter 7 discusses the potential of measuring individuals' adaptability and inventiveness, an important attribute for soldiers who routinely face unexpected and unique environments, situations, challenges, and opportunities. Adaptability involves the ability to adjust and accommodate to changing and unpredictable physical, interpersonal, cultural, and task environments. Inventiveness refers to the ability to think innovatively and produce novel high-quality and task-appropriate ideas, incorporating an orientation toward problem solving. Research on these two constructs suggests they increment predictive validity over other cognitive ability and personality measures for important outcomes such as performance and career continuance and progression.

Section 5. Methods and Methodology

This section includes three single chapters linked to the other research domains as methods and methodology for implementation.

Chapter 8. Psychometrics and Technology

Chapter 8 examines a variety of areas that show promise for improvements in measurement, including the application and modeling of forced-choice measurement methods, development of serious gaming, pursuit of Multidimensional Item Response Theory (MIRT), Big Data analytics, and other modern statistical tools. One example of the potential benefit is the likelihood that MIRT models can yield information about examinees' performance beyond what has been possible with traditional unidimensional IRT models. MIRT models may also offer improvements to test efficiency.

Chapter 9. Situations and Situational Judgment Tests

Chapter 9 focuses on the use of situations and situational judgment tests to measure and assess individuals' judgment abilities to interpret, evaluate, and weigh alternate courses of action appropriately and effectively. The chapter considers a variety of approaches to, and formats for, these tests; it discusses possible advantages of various presentation formats. For example, situational judgment tests administered in a video format may reduce the impact of lower verbal ability on test results and may provide a more immersive and engaging testing experience.

Chapter 10. Assessment of Individual Differences Through Neuroscience Measures

Chapter 10 examines neuroscience measures that may warrant consideration for testing applications in the near term, particularly as measures of anxiety, attention, and motivation in test takers. To illustrate, some level of anxiety is normal in test-taking situations. However, high levels of anxiety can have detrimental effects on test performance. Determining in real time through the use of the noninvasive technique of electroencephalography whether a candidate is experiencing such detrimental anxiety affords test administrators the opportunity to offer mitigation strategies to such candidates, thereby improving the degree to which assessment results offer an accurate representation of such candidates' abilities. Understanding candidates' levels of attention and motivation during testing can similarly yield better understanding of the credibility of test results.

REFERENCES

Armor, D., and P. Sackett. (2004). Manpower quality in the all-volunteer force. In B.A. Bicksler, C.L. Gilroy, and J.T. Warner, Eds., *The All-Volunteer Force: Thirty Years of Service* (pp. 90–108). Washington, DC: Brassey's.

Campbell, J.P. (2012). Behavior, performance and effectiveness in the twenty-first century. In S.W.J. Kozlowski, Ed., *The Oxford Handbook of Organizational Psychology, Volume 1* (pp. 159–194). New York: Oxford University Press.

Hough, L.M., and F.L. Oswald. (2000). Personnel selection: Looking toward the future—remembering the past. *Annual Review of Psychology*, 51(1):631–664.

National Research Council. (2013). *New Directions in Assessing Performance Potential of Individuals and Groups: Workshop Summary*. R. Pool, Rapporteur. Committee on Measuring Human Capabilities: Performance Potential of Individuals and Collectives. Board on Behavioral, Cognitive, and Sensory Sciences, Division of Behavioral and Social Sciences and Education. Washington, DC: The National Academies Press.

Nye, C.D., R. Su, J. Rounds, and F. Drasgow. (2012). Vocational interests and performance: A quantitative summary of over 60 years of research. *Perspectives on Psychological Science*, 7(4):384–403.

Rumsey, M.G., and J.M. Arabian. (2014). Military enlistment selection and classification: Moving forward. *Military Psychology*, 26(3):221–251.

Sackett, P.R., and F. Lievens. (2008). Personnel selection. *Annual Review of Psychology*, 59:419–450.

Sellman, W.S., D.H. Born, W.J. Strickland, and J.J. Ross. (2010). Selection and classification in the U.S. military. In J.L. Farr and N.T. Tippens, Eds., *Handbook of Employee Selection* (pp. 679–704). London, UK: Routledge.

Strickland, W.J., Ed. (2005). *A Longitudinal Examination of First Term Attrition and Reenlistment among FY 1999 Enlisted Accessions* (Technical Report 1172). Arlington, VA: U.S. Army Research Institute for the Behavioral and Social Sciences.

Trent, T., and J.H. Laurence. (1993). *Adaptability Screening for the Armed Forces*. Washington, DC: Office of the Assistant Secretary of Defense (Force Management and Personnel).
Van Iddekinge, C.H., P.L. Roth, D.J. Putka, and S.E. Lanivich. (2011). Are you interested? A meta-analysis of relations between vocational interests and employee performance and turnover. *Journal of Applied Psychology*, 96(6):1,167–1,194.

Section 2

Identification and Measurement of New Predictor Constructs

2

Fluid Intelligence, Working Memory Capacity, Executive Attention, and Inhibitory Control

Committee Conclusion: *The constructs of fluid intelligence (novel reasoning), working memory capacity, executive attention, and inhibitory control are important to a wide range of situations relevant to the military, from initial selection, selection for a particular job, and training regimes to issues having to do with emotional, behavioral, and impulse control in individuals after accession. These constructs reflect a range of cognitive, personality, and physiological dimensions that are largely unused in current assessment regimes. The committee concludes that these topics merit inclusion in a program of basic research with the long-term goal of improving the Army's enlisted accession system.*

The committee considers the areas of fluid intelligence, working memory capacity, executive attention, and inhibitory control as offering new constructs for the Army's consideration, even though some aspects of these ideas have been studied for decades. The newer research brings these several heretofore separate topics together and extends the relevance of the constructs beyond performance on specific tasks to broader issues of cognitive and emotional control. These topics are presented in a single chapter because there is considerable evidence that they overlap in terms of their theoretical motivations and definitions, their measurement, their variance, and their patterns of prediction. These topics are also brought together because, at the same time these overlaps are evident, future research must determine whether these various constructs reflect a single common mechanism or highly related but separate psychological mechanisms that might play different roles in the regulation of behavior, thought, and emotion.

If the latter hypothesis is supported, then a second issue is whether much more specific assessment of those separate mechanisms can add predictive validity for performance in the jobs for which potential military recruits are assessed.

Each section of the chapter begins with a brief history about one or more of the constructs listed in the title, focusing on how research on these constructs has converged and diverged over time. It then presents findings from various researchers who have studied these issues most recently, describes the evidence for the validity of the constructs in predicting performance of real-world tasks, and discusses the transition of what has been fairly basic research agenda on these topics to a more testing-oriented agenda. The sections end with a discussion of questions that should be addressed in future projects.

FLUID INTELLIGENCE

The idea that intelligence could be thought of as a general and therefore domain-free variable dates back at least to Spearman (1904). However, the idea that fluid and crystallized intelligence were separable was proposed by Spearman's student Raymond Cattell (1941) and elaborated by Cattell and his student John Horn (Horn and Cattell, 1966a, 1966b). As described in Cattell's biography by the website Human Intelligence:[1]

> Fluid abilities (*Gf*) drive the individual's ability to think and act quickly, solve novel problems, and encode short-term memories. They have been described as the source of intelligence that an individual uses when he or she doesn't already know what to do. Fluid intelligence is grounded in physiological efficiency, and is thus relatively independent of education and acculturation (Horn, 1967). The other factor, encompassing crystallized abilities (*Gc*), stems from learning and acculturation, and is reflected in tests of knowledge, general information, use of language (vocabulary) and a wide variety of acquired skills (Horn and Cattell, 1967). Personality factors, motivation and educational and cultural opportunity are central to its development, and it is only indirectly dependent on the physiological influences that mainly affect fluid abilities.

Fluid intelligence (*Gf*) is important for reasoning and novel problem solving, and there is strong and emerging evidence that it represents the heritable and biological aspect of intelligence (Plomin et al., 2008; Wright et al., 2007). Longitudinal and cross-sectional studies across the life span have repeatedly shown that, while crystallized intelligence—the culturally derived knowledge aspect of intelligence—remains high and even increases

---

[1] Available: http://www.intelltheory.com/rcattell.shtml [January 2015].

over the life span, G*f* declines over age (Horn and Cattell, 1967). In addition, individual differences in fluid intelligence (i.e., rank-order differences) appear to be quite stable over the life span (Deary et al., 2009, 2012). For example, Deary and his colleagues in the Lothian cohort studies made use of the fact that over 150,000 11-year-olds in the Lothian region of Scotland were tested for intelligence (IQ scores) more than 50 years ago and many of those individuals have been available for testing in recent years. Recently, Deary and colleagues (2012) conducted a genome-wide complex trait analysis on this sample and found a genetic correlation of 0.62 between intelligence in childhood and in old age. Furthermore, it appears that this relationship is higher for the lower quartile of abilities than for the upper quartile, which suggests that a more complete understanding of this relationship would be important for the selection and assignment of enlisted personnel.

The validity of fluid measures has been demonstrated for military-related tasks such as air traffic control (Ackerman and Cianciolo, 2002) and multitasking (Hambrick et al., 2010, 2011). The long-term stability and validity of fluid measures have been demonstrated in a sustained program of studies by David Lubinski and Camilla Benbow (2000, 2006). They started with a sample of 13-year-olds identified as being in the top 1 percent of individuals on measures of verbal and mathematical reasoning and tracked those individuals into middle adulthood (Lubinski and Benbow, 2006). Scores on these measures substantially predicted accomplishments in a wide array of domains in middle adulthood. Even at the highest levels, the scores obtained at age 13 predicted the number of patents, academic publications, and achievement in science and business at later ages.

The distinction between fluid and crystallized abilities becomes critically important in selection for the military. Recent papers have suggested that the Armed Services Vocational Aptitude Battery (ASVAB) is largely crystallized and that incremental validity can be added with measures of working memory capacity and fluid intelligence. The ASVAB does include a spatial ability subtest (Assembling Objects) which reflects a fluid ability in the typical examinee population (see Chapter 4, Spatial Abilities, for further discussion). Roberts and colleagues (2000) reported two studies, with a total of 7,100 subjects, showing that the ASVAB largely reflects acculturated learning and minimally reflects fluid abilities (G*f*). Hambrick and colleagues (2011) had Navy sailors perform a synthetic work task that simulated the multitasking demands of many different jobs. While the ASVAB did predict performance on this task, the ability to update working memory accounted for even more variance in the prediction of multitasking and synthetic work. Future research will be important to improve understanding of the mechanisms underlying fluid abilities and the differences between the mechanisms of working memory and fluid intelligence, including measures of these constructs as potential supplemental tests to the ASVAB.

There is ongoing military interest in and research on measures of fluid abilities. An expert panel charged with a review of the ASVAB recommended consideration of existing and new measures of fluid abilities as potential additions to the ASVAB (Drasgow et al., 2006). Alderton and colleagues (1997) examined a battery of tests in the spatial ability and working memory domains, administered in conjunction with the ASVAB. Their data show that Assembling Objects has a substantial loading on a general factor, as well as loading on a specific spatial ability factor. Thus, although it does indeed reflect a measure in the fluid abilities domain, it is likely not the best measure of fluid intelligence. Nonverbal reasoning tests, such as matrix tests, commonly produce very high general factor loadings, and a matrix test will be administered to all military applicants starting in April 2015 (see Russell et al., 2014).

The psychological and biological mechanisms reflected in standard tests of fluid intelligence and responsible for individual differences in the construct have been largely ignored in the psychometric literature and only recently have been addressed in the cognitive psychology and neuroscience literature. This lack of understanding of the specific cognitive abilities and the underlying biomarkers reflected in fluid intelligence is a gap in knowledge that it is important to fill to maximize the benefits of such assessments. If, for example, fluid intelligence is a composite of several underlying specific cognitive abilities it would be extremely useful to know whether those abilities are differentially related to various criterion measures and whether they might interact in some way that would be important to assess.

WORKING MEMORY CAPACITY AND EXECUTIVE ATTENTION

Measures of memory span (short-term memory) have been used to study memory abilities since Ebbinghaus (see Dempster, 1981). The first publication of a study using memory span as a measure (Jacobs, 1887) reported a strong relationship between a child's memory span and rank in class, and Francis Galton himself (1887) observed that few mentally deficient individuals could recall more than two items in a span test. Simple memory span tasks have been included in most large-scale tests of intelligence. Thus, from the beginning, what came to be called short-term memory appeared to reflect important individual differences in higher-order cognitive functions. The emergence of short-term memory as a major construct in cognitive psychology was predicated largely on research using span-like tasks, meaning that most of the work was done using serial recall of short lists of digits, letters, or words and with the same pool of items used over and over across lists. Crowder (1982), in a paper titled "The Demise of Short-term Memory," argued against two separate memory stores, and one of his arguments was based on the lack of relationship between measures of short-term memory and measures of real-world cognition. If short-term memory was important to real-world cognition, then individual differences in measures of that memory should correspond to individual differences in reading, learning, decision making, etc., and there was little evidence supporting that conclusion.

The picture clarified substantially when complex span measures were shown to have quite substantial correlations with reading and listening comprehension (Daneman and Carpenter, 1980; Engle and Kane, 2004). Examples of two complex spans alongside a simple letter span task, all of which require manipulation and remembering of verbal materials, are shown in Figure 2-1. In the reading span task, the subject is to read aloud the sentence and decide whether the sentence makes sense. That is followed by a letter to recall. In the operation span task, the subject is to calculate whether the equation is correct and then see a letter to recall. After two to seven such items, the subject is shown a set of question marks and asked to recall the to-be-remembered items.

Complex tasks may also involve the manipulation and remembering of nonverbal information such as the tasks in Figure 2-2. These tasks require the subject to make a decision about a pattern such as whether the rotated

Simple Span	Reading Span (WMC)	Operation Span (WMC)
B	The tiger leapt to the ridge. B	Is (3 x 1) − 1 = 3 ? B
N	I'll never forget my days of combat. N	Is (10 / 2) + 1 = 6 ? N
K	Andy was arrested for speeding. K	Is (8 / 4) − 1 = 1 ? K
J	The mirror cast a strange reflection. J	Is (3 x 3) + 1 = 12 ? J
S	Broccoli is a good source of nutrients. S	Is (4 x 3) + 2 = 14 ? S

FIGURE 2-1 Example of a simple span task, a reading span task, and an operation span task.
NOTE: WMC = working memory capacity.
SOURCE: Engle, Randall W. (2010). Role of working memory capacity in cognitive control. *Current Anthropology, 51*(S1):S17–S26. Reproduced by permission of and published by The University of Chicago Press.

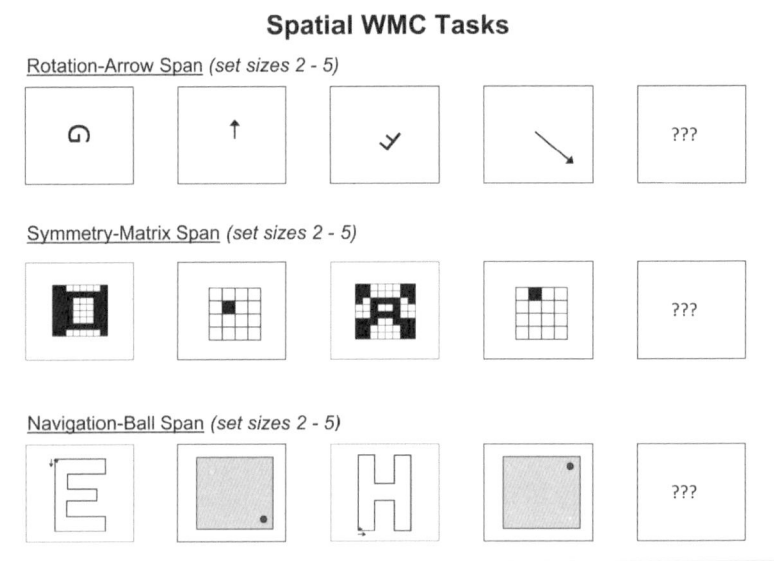

FIGURE 2-2 Three different spatial tasks.
NOTE: WMC = working memory capacity
SOURCE: Kane et al. (2004, p. 196).

block letter would be a correct letter when upright or whether the figure is symmetrical around a vertical axis. Each decision is followed by an item to be remembered such as the arrow pointing in one of eight directions and being one of two lengths, or a cell in a matrix.

One might think that tasks that differ as widely as those in Figures 2-1 and 2-2 would yield very different predictive validity for higher level tasks, but that is not the case. As shown in Figure 2-3, a huge array of such tasks has been shown to reflect a coherent latent factor. Further, that latent factor, typically called "working memory capacity" (WMC), has a very high relationship to the construct for fluid intelligence.

The wide array of WMC tasks have been shown to be quite valid in predicting performance on a huge variety of real-world cognitive tasks. Quoting from Engle and Kane (2004, p. 153):

> Scores on WMC tasks have been shown to predict a wide range of higher-order cognitive functions, including: reading and listening comprehension (Daneman and Carpenter, 1983), language comprehension (King and Just, 1991), following directions (Engle et al., 1991), vocabulary learning (Daneman and Green, 1986), note-taking (Kiewra and Benton, 1988), writing (Benton et al., 1984), reasoning (Barrouillet, 1996; Kyllonen and

Christal, 1990), bridge-playing (Clarkson-Smith and Hartley, 1990), and computer-language learning (Kyllonen and Stephens, 1990; Shute, 1991). Recent studies have begun to demonstrate the importance of WMC in the domains of social/emotional psychology and in psychopathology, either through individual-differences studies or studies using a working memory load during the performance of a task (Feldman-Barrett et al., in press [2004]). For example, low WMC individuals are less good at suppressing counterfactual thoughts, that is, those thoughts irrelevant to, or counter to, reality.

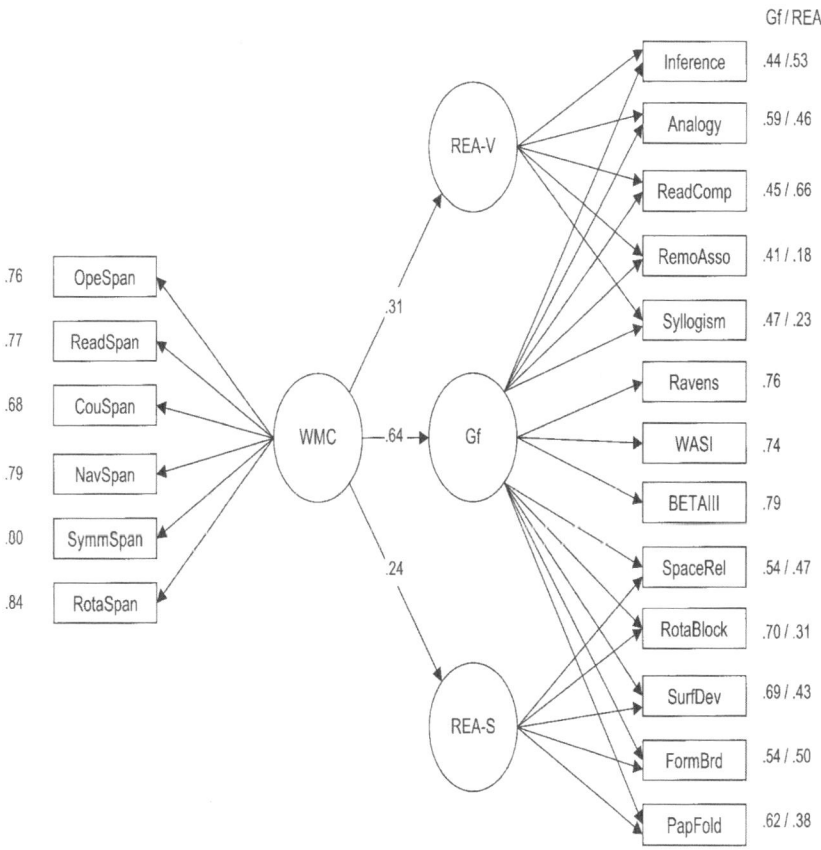

FIGURE 2-3 Path model for structural equation analysis of the relation between working memory capacity and reasoning factors.
SOURCE: Kane et al. (2004, p. 205).

The expert panel charged with a review of the ASVAB, described in the previous discussion of fluid abilities, also recommended consideration of working memory measures as potential additions to the ASVAB (Drasgow et al., 2006). Previously, Alderton and colleagues (1997) examined a battery of tests that included working memory measures, administered in conjunction with the ASVAB. Sager and colleagues (1997) offered evidence of the validity of working memory measures in this battery for predicting military training outcomes. Furthermore, a working memory test from this battery is currently being administered to Navy applicants (see Russell et al., 2014). Working memory measures were also explored in Project A, the Army's large-scale exploration of the relationship between a broad array of individual-differences constructs and various performance domains (Russell and Peterson, 2001; Russell et al., 2001).

Although the construct under discussion here is typically referred to as working memory capacity, there is strong and emerging evidence that the critical factor for regulation of thought and emotion is the ability to control one's attention, often referred to as executive attention (EA). EA refers to the ability to prevent attention capture by both endogenous and exogenous events (Engle and Kane, 2004). Individuals assessed to have lower EA are thought to be more likely to allow internally or externally generated events to capture their attention from tasks currently being performed. Thus, studies will often use the same tasks developed to measure WMC but will refer to the construct as Executive Attention.

There is a strong connection between the measures of WMC described above and measures of attention such as the Stroop task, antisaccade task, dichotic listening, and the flanker task. In an example of the antisaccade task, subjects stare at a fixation point on a computer screen while there are two boxes 11 degrees to each side of the fixation. At some point, one of the boxes will flicker and the subject is to look at the box on the *opposite* side of the screen. The flickering box affords movement, and evolution has predisposed us to look at that box since things that move have possible survival consequences. Performance can be measured either by eye movement analysis or by having the subject identify a briefly presented item in the box opposite to the flickering box (Kane et al., 2001; Unsworth et al., 2004); in both cases low WMC individuals are nearly twice as likely to make an error and glance at the flickering box. In the dichotic listening task, low WMC individuals are more than three times more likely than high WMC individuals to hear their name in the to-be-ignored ear.

The strong relationship of performance on these low-level attention tasks to the WMC tasks suggests that EA is likely to play a crucial role in both types of tasks. We do note that although EA is conceptualized as a cognitive ability, the pattern of relationships among various WMC tasks may also result from differences across participants in the degree of engagement with the tasks. Attributing relationships to EA differences alone requires the assumption of a common level of task motivation (usually a high level is assumed).

The concept of individual differences in WMC/EA has been used in explanations of psychopathologies such as alcoholism and schizophrenia. For example, Finn (2002) proposed a cognitive-motivational theory of vulnerability to alcoholism in which one key factor is WMC/EA. He argued that greater WMC allows an individual to better manipulate, monitor, and control the behavioral tendencies resulting from alcoholism, and that this directly affects the ability to resist a prepotent behavior such as taking a drink in spite of being aware that such behavior is ultimately maladaptive. Individual differences in WMC/EA have also been shown to be important in emotion regulation (Hofmann et al., 2011). Thus, assessment of whether individuals are likely to be more or less able to control impulses and self-destructive thoughts would benefit from inclusion of WMC measures.

The linkages between EA and impulse control suggest that examinations of EA may benefit from examining relations with self-control measures in the personality domain to determine the degree of overlap and potential incremental validity of one over the other. Recent studies have shown that the tendency to mind-wander during performance of a critical task is highly associated with measures of WMC (McVay and Kane, 2009, 2012a, 2012b). These researchers used a variety of techniques to measure what they called task-unrelated thoughts during performance of complex tasks. In one study (Kane et al., 2007), subjects carried a Palm Pilot[2] and were alerted eight random times over the course of their day to answer questions about the tasks they were currently performing, their level of concentration, how challenging the task was, how much effort they were expending, and whether their mind had wandered in the last few minutes. The results in Figure 2-4 show clearly that low and high WMC individuals differed greatly in their tendency to mind-wander and that the differences grew as more concentration was required in the task and the task became more challenging. Low WMC individuals are more likely to mind-wander as a task increases in challenge and effort level required. One question that could be investigated through future research would be the cause or effect related to whether mind wandering is a consequence of task difficulty and WMC or a predictor of WMC (suggesting that mind wandering is a consequence rather than a cause of WMC performance). These differences in performance would seem to be generalizable to a wide range of tasks performed in the Army across the full spectrum of operations from peacetime to combat situations.

---

[2] Palm Pilot was an early personal digital assistant that could be set up with multiple alarms and short interactive response-entry actions.

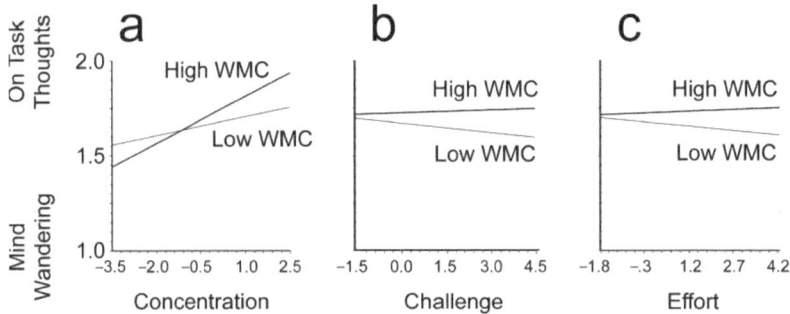

FIGURE 2-4 High versus low WMC individuals and task-unrelated thoughts in daily life.
NOTE: Values on the y-axis represent the mind wandering dependent variable, scored on each questionnaire as either 1 (for mind wandering) or 2 for on-task thoughts; lower values thus indicate more mind wandering. Values on the x-axis represent group-centered ratings for (a) concentration ("I had been trying to concentrate on what I was doing"), (b) challenge ("What I'm doing right now is challenging"), and (c) effort ("It takes a lot of effort to do this activity").
SOURCE: Kane, J.J., L.H. Brown, J.C. McVay, I. Myin-Germeys, P.J. Silva, and T.R. Kwapil. (2007). For whom the mind wanders, and when: An experience-sampling study of working memory and executive control in daily life. *Psychological Science, 18*(7):167. Reproduced by permission of SAGE Publications.

While a general mental abilities (i.e., Gf) approach is useful and has been considered the gold standard for predicting job performance (Schmidt and Hunter, 1998), recent work in this area suggests the importance of WMC in such predictions. In particular, WMC has been found to capture specific aptitudes beyond general mental abilities (Bosco and Allen, 2011; Hambrick et al., 2010; König et al., 2005). A recent study by König and colleagues (2005) testing 122 college students found that WMC was the best predictor of multitasking (similar conclusions were supported by Damos, 1993; Hambrick et al., 2010, 2011; and Stankov et al., 1989). These studies also showed WMC remained predictive of multitasking performance after controlling for fluid intelligence. In hierarchical regression analyses, WMC demonstrated the highest correlations with several measures of multitasking and predicted the most unique variance (Hambrick et al., 2010, 2011). Other research has found that WMC and Gf are distinct but strongly related (Kane et al., 2005).

Another perspective on assessments of WMC and EA is that, although they have great validity in predicting performance in real-world job situations, some research indicates they produce smaller mean racial/ethnic group differences than do measures of crystallized ability. Subgroup differences contribute to adverse impact, a violation of Title VII of the 1964 Civil Rights Act. Under that statute, a violation of Title VII[3] may be demonstrated by showing that an employment practice or policy has a disproportionately adverse effect on members of the protected class as compared with nonmembers of the protected class. Such impact is only acceptable to the extent that the practice is proven to be germane to the job being selected for. In other words, a test that has good validity and low adverse impact against a protected class is preferred over one that has good validity but has higher adverse impact.

A series of studies (Bosco and Allen, 2011) compared the EA battery developed by the Engle lab (Engle and Kane, 2004) with the Wonderlic test in terms of ability to predict job performance and associated adverse impact due to race (i.e., different mean scores for the two racial groups on the test). In three different studies, respectively involving college students, MBA students, and individuals working in a large financial firm, Bosco and Allen found that the EA battery accounted for greater variance in task or job performance than the Wonderlic test and had substantially less adverse impact. The EA battery predicted an additional 7.2 percent of the variance beyond the Wonderlic on the job simulation task, as well as an additional 5.2 percent of the variance in supervisor ratings of job performance. The reduced adverse impact for the EA battery was also found for supervisory ratings of managers in the workplace environment.

These findings are intriguing enough to mention; however, they are based on modest sample sizes, and additional replication is needed to solidify the basis of these findings. Verive and McDaniel (1996) report a meta-analysis of short-term memory tests on nearly 28,000 subjects and found that the black-white difference was less than half what it is on typical general cognitive ability tests, and yet the validity estimates remained high: .41 for job performance and .49 for training performance. Again, although interesting, the committee does not view these results as definitive. For example, the meta-analysis relies on untested assumptions about the degree of range restriction in the samples, and there is variance associated with these meta-analytic mean estimates that deserves to be understood.

Because short-term memory tests have been shown to be relatively unreliable and have reduced validity compared to measures of working memory capacity and executive attention (Engle et al., 1999a, 1999b), one might expect the latter measures to be even more resistant to adverse impact. This is consistent with recent work by Redick and colleagues (2012) in which gender differences were shown to be minimal on working memory complex span tasks over a sample size of 6,000 young adults.

---

[3] See 42 U.S.C. § 2000e-2. Available: http://www.eeoc.gov/laws/statutes/titlevii.cfm [December 2014].

Thus, the WMC/EA approach to assessment appears to provide substantial incremental validity for specific job situations and yet is less influenced by race or ethnic group. This tentative finding would seem to be particularly important for the modern Army situation but clearly needs further study and development, including research into cost-effective large-scale testing mechanisms suitable for administration in mobile or other non-laboratory settings without compromising validity, reliability, or test security. (See Section 5 of this report, Methods and Methodology, for further discussion of research topics to facilitate such developments.) In developing a future research program, it is important to recognize that although much research has been conducted on the constructs of fluid intelligence, WMC, and EA, research on the relationship between WMC and fluid intelligence is a relatively new and incomplete endeavor that combines two typically parallel research approaches: experimental and differential. Bringing these research approaches under one roof will improve the identification and understanding of the mechanisms responsible for the constructs of WMC, fluid intelligence, and EA, thus making significant contributions to the basic understanding of individual differences.

Research Recommendation:
Fluid Intelligence, Working Memory Capacity, and Executive Attention

The U.S. Army Research Institute for the Behavioral and Social Sciences should support research to understand the psychological, cognitive, and neurobiological mechanisms underlying the constructs of fluid intelligence (novel reasoning), working memory capacity, and executive attention.

A. Research should be conducted to ascertain whether these constructs reflect a common mechanism or are highly related but distinct mechanisms.
B. Assessments reflecting the results of research into the commonality versus distinctness of these constructs should be developed for purposes of validity investigations.
C. Ultimately, the basic research results from items A and B above should be used to inform research into time-efficient, computer-automated assessment(s).

INHIBITORY CONTROL

The research on WMC/EA described above illustrates how measures based on tasks conducted in the laboratory ("lab task measures") can be used to index individual differences in cognitive control or executive capacity that contribute to performance in various contexts. This body of cognitive-performance work represents an important extension of traditional personality-oriented research on variations in the tendency to restrain versus express impulses and emotions—research reflected in psychological constructs ranging from "ego control" (Block and Block, 1980) to "constraint" (Tellegen, 1985), "novelty seeking" (Cloninger, 1987), and "syndromes of disinhibition" (Gorenstein and Newman, 1980; Patterson and Newman, 1993). It would be useful to be able to predict with some accuracy those individuals who have difficulty controlling impulses for unacceptable behavior—that is, predicting cognitive, personality, and emotional characteristics that might lead to inappropriate or unacceptable behavior of the sort that has implications for an individual's military career or mission success.

Variations in performance on WMC/EA tasks and personality scale measures of impulsivity versus restraint can be viewed as indexing a common individual-differences construct. As evidence for this, capacities associated with WMC/EA appear to play a crucial role in the blocking or inhibition of intrusive thoughts (Brewin and Holmes, 2003). For example, individual differences in WMC are related to the ability to prevent unwanted information from intruding into consciousness and negatively affecting task performance. Individuals with greater measured WMC are better at suppressing unwanted thoughts when instructed to do so under experimental conditions, whether these thoughts are neutral (Brewin and Beaton, 2002) or obsessional (Brewin and Smart, 2005). These findings may help to explain why low intelligence, which is strongly correlated with WMC, is a risk factor for posttraumatic stress disorder (PTSD; Brewin et al., 2000). This relationship is particularly important to understand better, given the increasing number of members of the military reporting PTSD.

Other recent research (Patrick et al., 2012, 2013a, 2013b) indicates that assessment of inhibitory control can be extended to include physiological response measures, which may be of value for understanding processes underlying effective performance as well as adding to prediction of performance outcomes. Anterior brain structures, including regions of prefrontal cortex (Blumer and Benson, 1975; Damasio et al., 1990) and the anterior cingulate cortex, appear crucial for inhibitory control. The prefrontal cortex is theorized to be important for "top-down" processing, that is, guidance of behavior by internal representations of goals or states (Cohen and Servan-Schreiber, 1992; Miller, 1999; Wise et al., 1996). The anterior cingulate cortex has been conceptualized as a system that invokes the control functions of the prefrontal cortex as needed to successfully perform a task, either by detecting errors as they occur (Gehring et al., 1995; Scheffers et al., 1996), by monitoring conflict among competing response tendencies (Carter et al., 1998), or by estimating the likelihood of committing an error at the time a response is called for (Brown and Braver, 2005). Given the evidence for a brain basis to executive capacity, it should be possible to quantify individual differences in inhibitory control through brain response measures as well as through personality scale or lab performance measures.

However, some major challenges exist to incorporating physiological measures into assessment of individual-differences constructs like inhibitory control. In particular, while prominent models of personality include reference to neurobiological systems, the models themselves are based primarily on self-report personality data, with ideas about their connections to neurobiology formulated subsequently. As a consequence of this: (1) physiological variables tend to correlate only modestly with personality scale scores, as expected of measures from differing domains (cf. Campbell and Fiske, 1959), and (2) existing conceptions of individual differences tend to persist unaltered, rather than being reshaped by neurobiological findings.

A strategy for addressing these challenges as related to assessment of individual differences pertinent to performance in real-world contexts is the psychoneurometric approach (Patrick and Bernat, 2010; Patrick et al., 2012, 2013a, 2013b). This approach is grounded in classic perspectives on psychological assessment, which conceive of dispositional tendencies as constructs that transcend specific domains of measurement (Cronbach and Meehl, 1955; Loevinger, 1957). Viewed this way, ideas regarding the nature of a trait construct and how to measure it are considered provisional and subject to modification based on data.

Figure 2-5 depicts the psychoneurometric approach as applied to the individual-differences construct of inhibitory control, which can be operationalized psychometrically as disinhibition versus restraint (Krueger et al., 2007; see discussion below) or behaviorally (as discussed above) as performance on lab tasks that index cognitive control or executive capacity. The first step in the approach entails identifying reliable physiological indicators ($Phys_{var1}$, $Phys_{var2}$, etc., in Figure 2-5) of the target construct operationalized psychometrically—in the case of this illustration, as scores on a self-report measure of disinhibitory tendencies (i.e., disinhibition scale shown as $Cont_{DIS}$ in Figure 2-5). The next step entails mapping the interrelations among physiological variables known to correlate with the disinhibition scale measure to (1) establish a statistically reliable neurometric measure of inhibitory control (shown as $Cont_{neurometric}$ in Figure 2-5) and (2) develop understanding of brain circuits/processes associated with individual differences in inhibitory control. Knowledge gained about the convergence of multiple physiological indicators from different experimental tasks—and about brain mechanisms underlying this convergence—in turn feeds back into conceptualization and psychometric measurement of the target construct (large curved arrow on left side of Figure 2-5).

The Externalizing Spectrum Inventory (ESI) provides a comprehensive

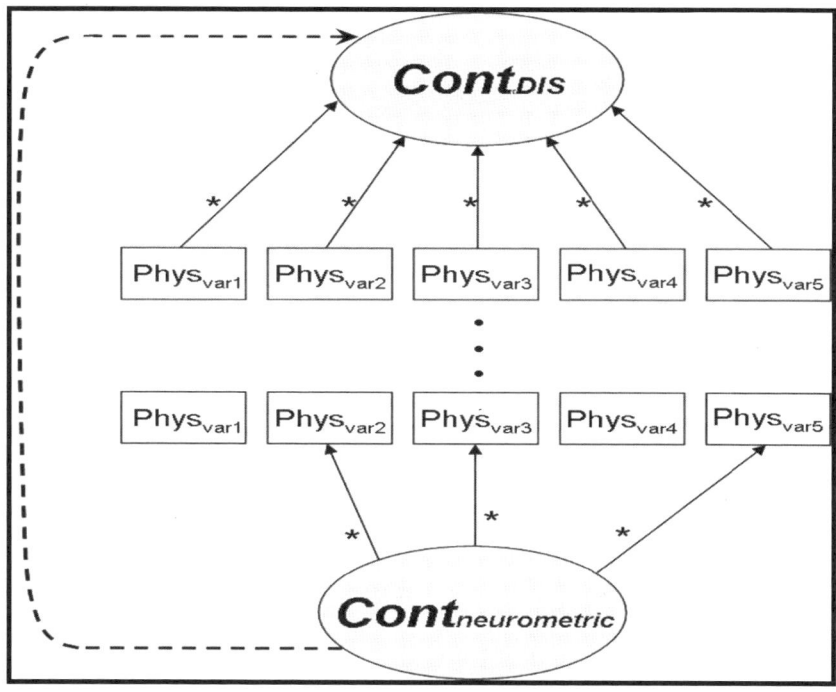

FIGURE 2-5 The psychoneurometric approach as applied to the individual-differences construct of inhibitory control.
NOTES: $Cont_{DIS}$ = construct of inhibitory control as assessed by self-report (i.e., disinhibition scale).
$Cont_{neurometric}$ = construct of inhibitory control as assessed by a composite of interrelated neurophysiological variables.
$Phys_{var}$ = physiological variable known to correlate reliably with inhibitory control as assessed by self-report.
SOURCE: Patrick, C.J., C.E. Durbin, and J.S. Moser. (2012). Reconceptualizing antisocial deviance in neurobehavioral terms. *Development and Psychopathology*, 24(3):1,064. Reproduced by permission of Cambridge University Press.

approach to assessing individual differences in inhibitory control through self-report (Krueger et al., 2007; Patrick et al., 2013a). It comprises 23 unidimensional subscales indexing tendencies toward impulsivity versus planful control, irresponsibility versus dependability, aggression in various forms versus empathic concern, fraudulence versus honesty, excitement seeking, rebelliousness and blame externalization, and use/abuse of alcohol and other drugs. As shown in Figure 2-6, the subscales of the ESI exhibit a

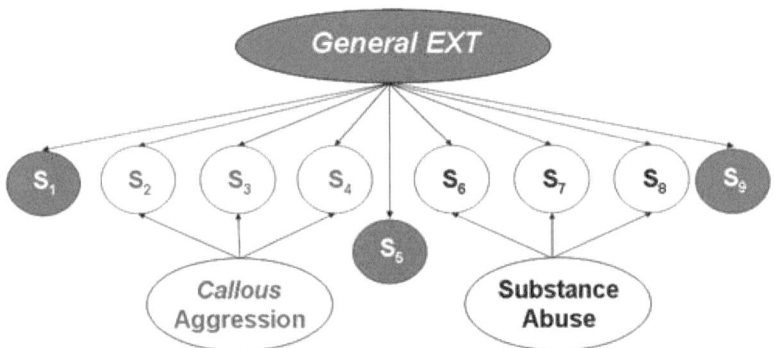

FIGURE 2-6 A schematic of the best fitting confirmatory bifactor model of the ESI (Krueger et al., 2007). The model is represented schematically because the 23 subscales of the ESI included in the model are too numerous to depict effectively in full.
NOTE: ESI = externalizing spectrum inventory; EXT = externalizing; S = scale, where the subscript numbers represent differing subscales.
SOURCE: Patrick, C.J., C.E. Durbin, and J.S. Moser. (2012). Reconceptualizing antisocial deviance in neurobehavioral terms. *Development and Psychopathology*, 24(3):1,050. Reproduced by permission of Cambridge University Press.

bifactor structure, with all scales loading on a general factor (*externalizing*, or disinhibition), and certain scales also loading on separate subfactors reflecting callous aggression and addiction proneness. Variations in general tendencies toward impulsiveness versus restraint associated with the broad disinhibition factor can be assessed using a brief scale consisting of 20 ESI items, referred to as DIS-20. This disinhibition scale does not include any aggression- or substance-related items from the ESI, but it nonetheless strongly *predicts* tendencies toward antisocial-aggressive behavior and substance problems (Patrick et al., 2012, 2013a, 2013b). It is not known whether the ESI will yield the same results in high-stakes testing situations, and the validity and susceptibility to faking or coaching is unknown.

The construct of inhibitory control has well-established brain correlates. According to Patrick and colleagues (2006), the best known indicator of this type is reduced amplitude of the P3 (or P300) brain potential response to task-relevant stimuli in the widely used 'oddball' task. They presented evidence that reduced P3 amplitude reflects general externalizing proneness (maladaptive acting out) as indexed by disorder symptoms. Differences between subjects high and low in disinhibition have also been shown for error-related negativity (ERN), the brain potential response that occurs when subjects make an error on cognitive tasks. Hall and colleagues (2007) demonstrated a negative relationship between amplitude of the ERN in a flanker task and levels of disinhibition as indexed by the ESI. This finding was replicated in subjects assessed for disinhibitory tendencies using the DIS-20 scale (Patrick et al., 2012); Figure 2-7 depicts average ERN waveforms for high versus low DIS-20 scorers based on a median split. Importantly, variations in inhibitory control assessed in these ways show correlations with lab task measures of executive capacity as well as with brain response measures. For example, in a study of twins, Young and colleagues (2009) reported a *genetic* correlation of -0.6 between disinhibitory tendencies as assessed by personality-trait and clinical-symptom measures and executive capacity as indexed by performance on WMC/EA tasks (i.e., heritable variance in disinhibitory tendencies was associated inversely, to a substantial degree, with heritable variance in executive capacity). Further-

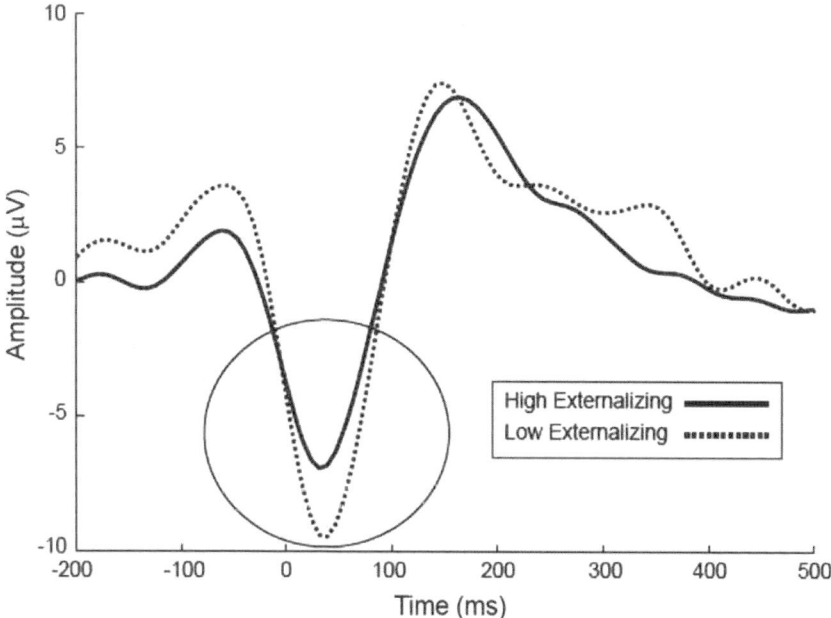

FIGURE 2-7 Mean error-related negativity (ERN) waveform for individuals high as compared to low in disinhibitory tendencies (i.e., above versus below the median on a 20-item disinhibition scale). The ERN (circled) reflects self-recognition of erroneous responses within a performance task (in this case, a speeded stimulus discrimination procedure).
SOURCE: Patrick, C.J., C.E. Durbin, and J.S. Moser. (2012). Reconceptualizing antisocial deviance in neurobehavioral terms. *Development and Psychopathology*, 24(3):1,057. Reproduced by permission of Cambridge University Press.

more, the likely overlap between inhibitory control capacity and individual differences in WMC/EA would be important to examine through future research to identify ways in which they are correlated or distinct.

Extending work on brain correlates of inhibitory control, Patrick and colleagues (2013b) demonstrated the effectiveness for predicting criterion variables across domains of clinical diagnosis (e.g., symptoms of antisocial and substance-related disorders) and neurophysiology (e.g., separate brain event-related potential [ERP] measures) of a composite psychometric-neurophysiological (psychoneurometric) index of trait disinhibition. This composite index consists of two brain-ERP indicators and scores on the DIS-20 disinhibition scale, along with another self-report measure of trait disinhibition. The psychoneurometric index was developed using data from one large participant sample ($N = 393$) and evaluated for predictive validity in a separate cross-validation sample ($N = 60$). Figure 2-8 depicts results for the cross-validation sample. The purple bars (with their tops circled) represent the correlations between scores on the four-indicator psychoneurometric (disinhibition-scale/brain-ERP) factor and criterion variables consisting of (1) a composite of separate ERP variables (i.e., target stimulus P3 from an oddball task, feedback stimulus P3 from a choice-feedback task, and response-locked ERN from a flanker task) and (2) symptoms of differing impulse-control disorders as assessed by clinical interview. Depicted in the figure for purposes of comparison are correlations for the ESI Disinhibition scale indicator of the DIS/ERP factor alone (gray bars) and the mean of the two ERP indicators alone with the composite ERP and diagnostic criterion variables (pink bars). A minus sign (–) above certain bars denotes a negative correlation coefficient for the variable indicated.

The data summarized by Figure 2-8 show that the psychoneurometric factor predicted criterion variables in the diagnostic and brain response domains to comparable robust degrees: the correlations for this factor with ERP composite scores and diagnostic composite scores (purple bars) both exceeded 0.6. By contrast, ESI-Disinhibition scores alone (gray bars) predicted criterion variables in the diagnostic domain very effectively but predicted criteria in the brain response domain only modestly. The ERP indicators alone (pink bars) predicted criterion variables in the brain response domain very effectively but predicted criteria in the diagnostic domain only modestly.

Research Recommendation: Inhibitory Control

The U.S. Army Research Institute for the Behavioral and Social Sciences should support research to further understanding of inhibitory control, including but not limited to the following lines of inquiry:

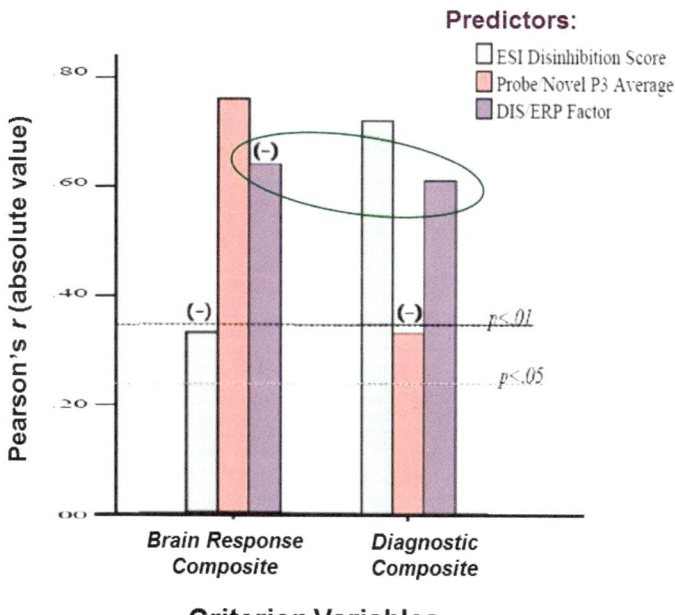

FIGURE 2-8 Associations with independent composite indices of brain response (left bars) and diagnostic symptoms (right bars) for three measures of disinhibitory tendencies: (1) scores on a 20-item disinhibition (DIS) scale (gray bars), (2) mean of two P3 brain responses (ERP) indicators of disinhibition (pink bars), and (3) composite of two self-report and two P3 brain indicators of disinhibition (purple bars). (-) = direction of correlational association is negative. Similar magnitude of rs for DIS/ERP predictor with brain and diagnostic criteria (circled purple bars) indicates that this psychoneurometric measure predicts effectively across these two domains of measurement.
SOURCE: Patrick et al. (2013b, p. 913).

A. Develop time-efficient, computer-automated self-report and behavioral assessments of inhibitory control capacity that demonstrate convergence with neurophysiological indices, as well as differentiation from constructs considered distinct from inhibitory control.
B. Examine the extent to which inhibitory control—as assessed through self-report, task-behavioral, and physiological response measures—predicts performance outcomes of interest (e.g., accidents, disciplinary incidents) and understand the common and unique aspects of the different assessment approaches in terms of underlying processes tapped by each and how these processes relate to performance.

REFERENCES

Ackerman, P.L., and A.T. Cianciolo. (2002). Ability and task constraint determinants of complex task performance. *Journal of Experimental Psychology: Applied, 8*(3):194–208.
Alderton, D.L., J.H. Wolfe, and G.E. Larson. (1997). The ECAT battery. *Military Psychology, 9*(1):5–37.
Barrouillet, P. (1996). Transitive inferences from set-inclusion relations and working memory. *Journal of Experimental Psychology: Learning, Memory, and Cognition, 22*(6):1,408–1,422.
Benton, S.L., R.G. Kraft, J.A. Glover, and B.S. Plake. (1984). Cognitive capacity differences among writers. *Journal of Educational Psychology, 76*(5):820–834.
Block, J., and J.H. Block. (1980). The role of ego-resiliency and ego-control in the organization of behavior. In W.A. Collins, Ed., *Minnesota Symposium on Child Psychology* (pp. 39–101). Hillsdale, NJ: Erlbaum.
Blumer, D., and D. Benson. (1975). Personality changes with frontal and temporal lesions. In D.F. Benson and F. Blumer, Eds., *Psychiatric Aspects of Neurologic Disease* (pp. 151–170). New York: Grune and Stratton.
Bosco, F.A., and D.G. Allen. (August 2011). Executive attention as a predictor of employee performance. *Academy of Management Annual Meeting Proceedings, 2011*(1):1–6.
Brewin, C.R., and A. Beaton. (2002). Thought suppression, intelligence, and working memory capacity. *Behaviour Research and Therapy, 40*(8):923–930.
Brewin, C.R., and E.A. Holmes. (2003). Psychological theories of posttraumatic stress disorder. *Clinical Psychology Review, 23*(3):339–376.
Brewin, C.R., and L. Smart. (2005). Working memory capacity and suppression of intrusive thoughts. *Journal of Behavior Therapy and Experimental Psychiatry, 36*(1):61–68.
Brewin, C.R., B. Andrews, and J.D. Valentine. (2000). Meta-analysis of risk factors for posttraumatic stress disorder in trauma-exposed adults. *Journal of Consulting and Clinical Psychology, 68*(5):748–766.
Brown, J., and T. Braver. (2005). Learned predictions of error likelihood in the Anterior Cingulate Cortex. *Science, 307*(5,712):1,118–1,121.
Campbell, D.T., and D.W. Fiske. (1959). Convergent and discriminant validation by the multitrait-multimethod matrix. *Psychological Bulletin, 56*(2):81–105.
Carter, C.S., T.S. Braver, D.M. Barch, M.M. Botvinick, D.N. Noll, and J.D. Cohen. (1998). Anterior cingulate cortex, error detection, and the online monitoring of performance. *Science, 280*(5,364):747–749.
Cattell, R.B. (1941). Some theoretical issues in adult intelligence testing. *Psychological Bulletin, 38*:592.
Clarkson-Smith, L., and A.A. Hartley. (1990). The game of bridge as an exercise in working memory and reasoning. *Journal of Gerontology, 45*(6):233–238.
Cloninger, C.R. (1987). A systematic method for clinical description and classification of personality variants. *Archives of General Psychiatry, 44*(6):573–588.
Cohen, J.D., and D. Servan-Schreiber. (1992). Context, cortex, and dopamine: A connectionist approach to behavior and biology in schizophrenia. *Psychological Review, 99*(1):45–77.
Cronbach, L.J., and P.E. Meehl. (1955). Construct validity in psychological tests. *Psychological Bulletin, 52*(4):281–302.
Crowder, R.G. (1982). The demise of short term memory. *Acta Psychologica, 50*(3):291–323.
Damos, D.L. (1993). Using meta-analysis to compare the predictive validity of single- and multiple-task measures of flight performance. *Human Factors, 35*(4):615–628.
Damasio, A.R., D. Tranel, and H. Damasio. (1990). Individuals with sociopathic behavior caused by frontal damage fail to respond automatically to social stimuli. *Behavioral Brain Research, 41*(2):81–94.

Daneman, M., and P.A. Carpenter. (1980). Individual differences in working memory and reading. *Journal of Verbal Learning and Verbal Behavior,* 19(4):450–466.

Daneman, M., and P.A. Carpenter. (1983). Individual differences in integrating information between and within sentences. *Journal of Experimental Psychology: Learning, Memory, and Cognition,* 9(4):561–584.

Daneman, M., and I. Green. (1986). Individual differences in comprehending and producing words in context. *Journal of Memory and Language,* 25(1):1–18.

Deary, I.J., L.J. Whalley, and J.M. Starr. (2009). *A Lifetime of Intelligence: Follow-up studies of the Scottish mental surveys of 1931 and 1947.* Washington, DC: American Psychological Association.

Deary, I.J., J. Yang, G. Davies, S.E. Harris, A. Tenesa, D. Liewald, M. Luciano, L.M. Lopez, A.J. Gow, J. Corley, P. Redmond, H.C. Fox, S.J. Rowe, P. Haggarty, G. McNeill, M.E. Goddard, D.J. Porteous, L.J. Whalley, J.M. Starr, and P.M. Visscher. (2012). Genetic contributions to stability and change in intelligence from childhood to old age. *Nature,* 482(7,384):212–215.

Dempster, F.N. (1981). Memory span: Sources of individual and developmental differences. *Psychological Bulletin,* 89(1):63–100.

Drasgow, F., S.E. Embretson, P.C. Kyllonen, and N. Schmitt. (2006). *Technical Review of the Armed Services Vocational Aptitude Battery (ASVAB) (FR-06-25).* Alexandria, VA: Human Resources Research Organization.

Engle, R.W. (2010). Role of working-memory capacity in cognitive control. *Current Anthropology,* 51(S1):S17–S26.

Engle, R.W., and M.J. Kane. (2004). Executive attention, working memory capacity, and a two-factor theory of cognitive control. In B. Ross, Ed., *The Psychology of Learning and Motivation, Vol. 44* (pp. 145–199). New York: Academic Press.

Engle, R.W., J.J. Carullo, and K.W. Collins. (1991). Individual differences in working memory for comprehension and following directions. *Journal of Educational Research,* 84(5):253–262.

Engle, R.W., S.W. Tuholski, J.E. Laughlin, and A.R.R. Conway. (1999a). Working memory, short-term memory and general fluid intelligence: A latent variable approach. *Journal of Experimental Psychology: General,* 128(3):309–331.

Engle, R.W., M.J. Kane, and S.W. Tuholski. (1999b). Individual differences in working memory capacity and what they tell us about controlled attention, general fluid intelligence and functions of the prefrontal cortex. In A. Miyake and P. Shah, Eds., *Models of Working Memory: Mechanisms of Active Maintenance and Executive Control* (pp. 102–134). London, UK: Cambridge University Press.

Feldman-Barrett, L., M.M. Tugade, and R.W. Engle. (2004). Individual differences in working memory capacity and dual-process theories of the mind. *Psychological Bulletin,* 130(4):553–573.

Finn, P.R. (2002). Motivation, working memory, and decision making: A cognitive-motivational theory of personality vulnerability to alcoholism. *Behavioral and Cognitive Neuroscience Reviews,* 3:183–205.

Galton, F. (1887). Supplementary notes on prehension in idiots. *Mind,* 12(45):79–82.

Gehring W.J., M.G.H. Coles, D.E. Meyer, and E. Donchin. (1995). A brain potential manifestation of error-related processing. In G. Karmos, M. Molnar, V. Csepe, I. Czigler, and J.E. Desmedt, Eds., *Perspectives of Event-Related Potentials Research* (pp. 267–272). Amsterdam, the Netherlands: Elsevier.

Gorenstein, E.E., and J.P. Newman (1980). Disinhibitory psychopathology: A new perspective and a model for research. *Psychological Review,* 87(3):301–315.

Hall, J.R., E.M. Bernat, and C.J. Patrick. (2007). Externalizing psychopathology and the error-related negativity. *Psychological Science,* 18(4):326–333.

Hambrick, D.Z., F.L. Oswald, E.S. Darowski, T.A. Rench, and R. Brou. (2010). Predictors of multitasking performance in a synthetic work paradigm. *Applied Cognitive Psychology,* 24(8):1,149–1,167.
Hambrick, D.Z., T.A. Rench, E.M. Poposki, E.S. Darowski, D. Roland, R.M. Bearden, F.L. Oswald, and R. Brou. (2011). The relationship between the ASVAB and multitasking in Navy sailors: A process-specific approach. *Military Psychology,* 23(4):365–380.
Hofmann, W., M. Friese, B.J. Schmeichel, and A.D. Baddeley. (2011). Working memory and self-regulation. In K.D. Vohs and R.F. Baumeister, Eds., *Handbook of Self-Regulation: Research, Theory, and Applications, 2nd Edition* (pp. 204–225). New York: Guilford Press.
Horn, J.L. (1967). Intelligence: Why it grows, why it declines. *Society,* 5(1):23–31.
Horn, J.L., and R.B. Cattell. (1966a). Refinement and test of the theory of fluid and crystallized general intelligences. *Journal of Educational Psychology,* 57(5):253–270.
Horn, J.L., and R.B. Cattell. (1966b). Age differences in primary mental ability factors. *Journal of Gerontology,* 21(2):210–220.
Horn, J.L., and R.B. Cattell. (1967). Age differences in fluid and crystallized intelligence. *Acta Psychologica,* 26:107–129.
Jacobs, J. (1887). Experiments on "prehension." *Mind,* 12(45):75–79.
Kane, M.J., K.M. Bleckley, A.R.A.Conway, and R.W. Engle. (2001). A controlled-attention view of working-memory capacity. *Journal of Experimental Psychology: General,* 130:169–183.
Kane, M.J., D.Z. Hambrick, S.W. Tuholski, O. Wilhelm, T.W. Payne, and R.W. Engle. (2004). The generality of working-memory capacity: A latent-variable approach to verbal and visuo-spatial memory span and reasoning. *Journal of Experimental Psychology: General* 133(2):189–217.
Kane, M.J., D.Z. Hambrick, and A.R.A. Conway. (2005). Working memory capacity and fluid intelligence are strongly related constructs: Comment on Ackerman, Beier, and Boyle. *Psychological Bulletin,* 131(1):66–71.
Kane, M.J., L.H. Brown, J.C. McVay, I. Myin-Germeys, P.J. Silvia, and T.R. Kwapil. (2007). For whom the mind wanders, and when: An experience-sampling study of working memory and executive control in daily life. *Psychological Science,* 18(7):614–621.
Kiewra, K.A., and S.L. Benton. (1988). The relationship between information processing ability and notetaking. *Contemporary Educational Psychology.* 13(1):33–44.
King, J., and M.A. Just. (1991). Individual differences in syntactic processing: The role of working memory. *Journal of Memory and Language* 30(5):580–602.
Kyllonen, P.C., and R.E. Christal. (1990). Reasoning ability is (little more than) working-memory capacity? *Intelligence,* 14(4):389–433.
Kyllonen, P.C., and D.L. Stephens. (1990). Cognitive abilities as determinants of success in acquiring logic skill. *Learning and Individual Differences,* 2(2):129–160.
König, C.J., M. Buhner, and G. Murling (2005). Working memory, fluid intelligence, and attention are predictors of multitasking performance, but polychronicity and extraversion are not. *Human Performance,* 18(3):243–266.
Krueger, R.F., K.E. Markon, C.J. Patrick, S.D. Benning, and M.D. Kramer. (2007). Linking antisocial behavior, substance use, and personality: An integrative quantitative model of the adult externalizing spectrum. *Journal of Abnormal Psychology,* 116(4):645–666.
Loevinger, J. (1957). Objective tests as instruments of psychological theory. Monograph supplement 9. *Psychological Reports,* 3:635–694.
Lubinski, D., and C.P. Benbow. (2000). States of excellence. *The American Psychologist,* 55(1):137–150.

Lubinski, D., and C.P. Benbow. (2006). Study of mathematically precocious youth after 35 years: Uncovering antecedents for the development of math-science expertise. *Perspectives on Psychological Science, 1*(4):316–345.

McVay, J.C., and M.J. Kane. (2009). Conducting the train of thought: Working memory capacity, goal neglect, and mind wandering in an executive-control task. *Journal of Experimental Psychology: Learning, Memory, and Cognition, 35*(1):196–204.

McVay, J.C., and M.J. Kane. (2012a). Drifting from slow to "D'oh!" Working memory capacity and mind wandering predict extreme reaction times and executive-control errors. *Journal of Experimental Psychology: Learning, Memory, and Cognition, 38*(3):525–549.

McVay, J.C., and M.J. Kane. (2012b). Why does working memory capacity predict variation in reading comprehension? On the influence of mind wandering and executive attention. *Journal of Experimental Psychology: General, 141*(2):302–320.

Miller, E.K. (1999). The prefrontal cortex: Complex neural properties for complex behavioral. *Neuron, 22*(1):15–17.

Patrick, C.J., and E.M. Bernat. (2010). Neuroscientific foundations of psychopathology. In T. Millon, R.F. Krueger, and E. Simonsen, Eds., *Contemporary Directions in Psychopathology: Toward the DSM-V* (pp. 419–452). New York: Guilford Press.

Patrick, C.J., E.M. Bernat, S.M. Malone, W.G. Iacono, R.F. Krueger, and M.K. McGue. (2006). P300 amplitude as an indicator of externalizing in adolescent males. *Psychophysiology, 43*(1):84–92.

Patrick, C.J., C.E. Durbin, and J.S. Moser. (2012). Reconceptualizing proneness to antisocial deviance in neurobehavioral terms. *Development and Psychopathology, 24*(3):1,047–1,071.

Patrick, C.J., M.D. Kramer, R.F. Krueger, and K.E. Markon. (2013a). Optimizing efficiency of psychopathology assessment through quantitative modeling: Development of a brief form of the Externalizing Spectrum Inventory. *Psychological Assessment, 25*(4):1,332–1,348.

Patrick, C.J., N.C. Venables, J.R. Yancey, B.M. Hicks, L.D. Nelson, and M.D. Kramer. (2013b). A construct-network approach to bridging diagnostic and physiological domains: Application to assessment of externalizing psychopathology. *Journal of Abnormal Psychology, 122*(3):902–916.

Patterson, C.M., and J.P. Newman. (1993). Reflectivity and learning from aversive events: toward a psychological mechanism for the syndromes of disinhibition. *Psychological Review, 100*(4):716–736.

Plomin, R., J.C. Defries, G.E. McClearn, and P. McGuffin. (2008). *Behavioral Genetics.* (5th Ed.). New York: Worth.

Roberts, R.D., G.N. Goff, F. Anjoul, P.C. Kyllonen, G. Pallier, and L. Stankov. (2000). The Armed Services Vocational Aptitude Battery (ASVAB): Little more than acculturated learning (Gc)!? *Learning and Individual Differences, 12*(1):81–103.

Redick, T.S., J.M. Broadway, M.E. Meier, P.S. Kuriakose, N. Unsworth, M.J. Kane, and R.W. Engle. (2012). Measuring working memory capacity with automated complex span tasks. *European Journal of Psychological Assessment, 28*(3):164–171.

Russell, T.L., and N.G. Peterson. (2001). The experimental battery: Basic attribute scores for predicting performance in a population of jobs. In J.P. Campbell and D.J. Knapp, Eds., *Exploring the Limits in Personnel Selection and Classification* (pp. 269–306). Hillsdale, NJ: Lawrence Erlbaum.

Russell, T.L., N.G. Peterson, R.L. Rosse, J.T. Hatten, J.J. McHenry, and J.S. Houston. (2001). The measurement of cognitive, perceptual and psychomotor abilities. In J.P. Campbell and D.J. Knapp, Eds., *Exploring the Limits in Personnel Selection and Classification* (pp. 71–110). Hillsdale, NJ: Lawrence Erlbaum.

Russell, T.L., L. Ford, and P. Ramsberger, Eds. (2014). *Thoughts on the Future of Military Enlisted Selection and Classification.* Alexandria, VA: Human Resource Research Organization.

Sager, C.E., N.G. Peterson, S.H. Oppler, R.L. Rosse, and C.B. Walker. (1997). An examination of five indexes of test battery performance: Analysis of the ECAT battery. *Military Psychology*, 9(1):97–120.

Scheffers, M., M.G.H. Coles, P. Bernstein, W. Gehring, and E. Donchin. (1996). Event-related brain potentials and error-related processing: An analysis of incorrect responses to go and no-go stimuli. *Psychophysiology*, 33(1):42–53.

Schmidt, F.L., and J.E. Hunter. (1998). The validity and utility of selection methods in personnel psychology: Practical and theoretical implications of 85 years of research findings. *Psychological Bulletin*, 124(2):262–274.

Shute, V.J. (1991). Who is likely to acquire programming skills? *Journal of Education Computing Research*, 7(1):1–24.

Spearman, C. (1904) "General intelligence," objectively determined and measured. *The American Journal of Psychology*, 15(2):201–292.

Stankov, L., G. Fogarty, and C. Watt. (1989). Competing tasks: Predictors of managerial potential. *Personality and Individual Differences*, 10(3):295–302.

Tellegen, A. (1985). Structures of mood and personality and their relevance to assessing anxiety, with an emphasis on self-report. In A.H. Tuma and J.D. Maser, Eds., *Anxiety and the Anxiety Disorders* (pp. 681–706). Hillsdale, NJ: Lawrence Erlbaum.

Unsworth, N., J.C. Schrock, and R.W. Engle. (2004). Working memory capacity and the antisaccade task: Individual differences in voluntary saccade control. *Journal of Experimental Psychology: Learning, Memory, and Cognition*, 30:1,302–1,321.

Verive, J.M., and M.A. McDaniel. (1996). Short-term memory tests in personnel selection: Low adverse impact and high validity. *Intelligence*, 23(1):15–32.

Wise, S.P., E.A. Murray, and C.R. Gerfen. (1996). The frontal-basal ganglia system in primates. *Critical Reviews in Neurobiology*, 10(3-4):317–356.

Wright, S.B., B.J. Matlen, C.L. Baym, E. Ferrer, and S.A. Bunge. (2007). Neural correlates of fluid reasoning in children and adults. *Frontiers in Human Neuroscience*, 1:8. Available: http://www.ncbi.nlm.nih.gov/pmc/articles/PMC2525981/ [September 2014].

Young, S. E., N.P. Friedman, A. Miyake, E.G. Willcutt, R.P. Corley, B.C. Haberstick, and J.K. Hewitt. (2009). Behavioral disinhibition: Liability for externalizing spectrum disorders and its genetic and environmental relation to response inhibition across adolescence. *Journal of Abnormal Psychology*, 118(1):117–130.

3

Cognitive Biases

Committee Conclusion: Cognitive biases, such as confirmation bias, anchoring, overconfidence, sunk cost, availability, and others, appear broadly relevant to the military because of findings, from both the analysis of large-scale disasters and the broader literature on cognitive biases, that show how irrational decision making results from failing to reflect on choices. Research on a tendency to engage in cognitive biases as a stable individual-differences measure is limited, and there are measurement challenges that must be dealt with before operational cognitive bias assessment could be implemented. The conceptual relevance of this topic, paired with the limited research to date, which takes an individual-differences orientation, leads the committee to conclude that cognitive biases merit inclusion in a program of basic research with the long-term goal of improving the Army's enlisted accession system.

Decision biases or cognitive biases refer to ways of thinking or a thought process that produces errors in judgment or decision making, or at least departures from the use of normative rules or standards (Gilovich and Griffin, 2002). A prevailing model is that cognitive biases result from the use of thinking shortcuts or heuristics, where such shortcuts lead to wrong decisions (Tversky and Kahneman, 1974). Not all thinking shortcuts, or heuristics, lead to wrong or poor decisions; in fact they can lead to good decisions in many contexts, and in some contexts they can lead to better decisions than those given by more deliberate approaches (e.g., Gigerenzer et al, 2011; Vickrey et al., 2010). Nevertheless in many circumstances cognitive biases can lead to poor decisions. In such cases, the thinking associated with cognitive biases is often assumed to be fast, nonconscious, automatic, not requiring working memory resources, and independent of cognitive ability. It is sometimes referred to as System 1 thinking, in contrast to the more careful, controlled, memory-dependent, rule-based, correlated with cognitive ability, and deliberate System 2 thinking (Evans and Stanovich, 2003; Kahneman, 2011).

EXAMPLES OF COGNITIVE BIASES IN ACTION

An example of the kind of irrationality in thinking and judgment produced by cognitive biases was described by Ariely (2008). He conducted an experiment based on a magazine advertising campaign offering choices of $59 for Internet-only subscriptions, $125 for print-only, and $125 for Internet-plus-print subscriptions. The latter seems like the best deal because it seems to offer the Internet access for free, and most people in the experiment took it. But if not given the print-only option, people were twice as likely to choose the Internet option. Ariely's experiment demonstrates how decision making is influenced by relative advantages of one option over another. The print-only option was a decoy, presented only to make the $125 combination offer more attractive. No one chose the print-only option, but it affected people's choices between the other two options. In many military decision-making contexts (e.g., how to approach a target, who is judged to be friend or foe), cognitive biases may influence the quality of the decisions and their outcomes.

Disasters and Tragedies

Many reports of major disasters invoke cognitive biases as at least partly responsible for errors in judgment that may have led to the disaster. For example, in 1996, eight mountain climbers died on Mt. Everest when a snowstorm caught them near the summit. Roberto (2002) reviewed accounts from surviving climbers and suggested that three cognitive biases may be partly responsible for the tragedy. The sunk cost effect may have occurred when climbers insisted on continuing to the summit after expending much time and energy on the ascent. The escalated commitment to get to the top meant that insufficient resources were left for a safe descent during the storm. Second, two expedition leaders may have been overconfident in their skills, biasing their judgments and risk assessments to bring their clients to the summit. Third, past expeditions were conducted under good weather conditions. A recency bias may have contributed to the leaders' overconfidence and underestimation of the dangers from a storm. Unfortunately, the poor decisions by these two leaders led to their deaths and the deaths of three of their team members.

Consider the Iran Air Flight 655 incident in which an Iran Air civilian passenger flight was shot down by surface-to-air missiles fired from the USS *Vincennes* over the Persian Gulf, killing all 290 passengers on board. The commanding officer had incorrectly acted upon the belief that the Iranian Airbus was actually an F-14 fighter from the Iranian Air Force; a belief developed in the context of a high pressure situation with complicated, confusing, and contradictory information to be interpreted and reconciled within minutes (U.S. Department of Defense, 1988). As tragedies like this often go, there were many factors that contributed to the mistake (U.S. Department of Defense, 1988). However, a possible contributing cause is certainly confirmation bias, in which the context of high tensions and prior incidents in the area (including the 1987 incident in which an Iraqi jet determined to be nonhostile shot upon the USS *Stark*, killing 37 sailors and injuring 21 more) contributed to confirmatory thinking such that the evidence of a military aircraft was overweighted compared to the disconfirmatory evidence of a civilian aircraft.

Confirmation biases negatively affect decisions when individuals interpret information, including conflicting evidence, as confirmation of previously held beliefs. This is a tendency of special concern in situations where information is incomplete or unclear and critical decisions must be made under high levels of uncertainty, such as decisions that must be made in combat or by intelligence analysts (see Spellman, 2011, for further discussion of individual reasoning applied to the tasks of intelligence analysts). The detrimental effects of confirmation bias are well known to the Intelligence Community, and many tools and techniques have been developed to assist intelligence analysts in avoiding them (Heuer, 1999; Heuer and Pherson, 2011).

Cognitive Biases in Everyday Reasoning and Decision Making

Besides the role cognitive biases might play in well-known tragedies and disasters, cognitive biases may routinely enter into everyday decision making and may be particularly important in military contexts. For example, soldiers are often put into the position of having to judge others' motives, such as having to judge the motives of host-nation citizens or international coalition military members. Cognitive biases such as projection (assuming others share our own feelings, attitudes, and values) can distort such judgments. Judging whether another person is friend or foe can be influenced by various cognitive biases. Humans are often poor judges of current and future events; for example, people often assume that someone or something can be adequately categorized on the basis of a single feature, such as an article of clothing or a head covering. An example is the murder of Balbir Singh Sodhi, a Sikh and a gas station owner in Mesa, Arizona, shortly after the September 11, 2001, terrorist attacks in the United States (Lewin, 2001). Or people jump to a conclusion about someone else's motives and assume that the other person is doing something because of "the way they are"—their culture or personality—without taking into account a more local and specific reason for the action. This is an example of fundamental attribution error: the tendency to attribute others' mistakes to something about *them* and to attribute our own mistakes to something external to ourselves (Ross, 1977). Cognitive biases can also creep into ratings—for example, our first impressions of something or someone might lead to a hard-to-alter belief about that thing or person due to confirmation bias, to fundamental attribution error, to anchoring (the tendency to place undue value on the first pieces of information received), or to representativeness (the perception of similarity between objects and comparison to a prototype; see Tversky and Kahneman, 1974). Cognitive biases are often invoked in explanations for failures "to connect the dots" and for failures of sensitivity to cultural differences.

These examples suggest that cognitive biases operate in everyday reasoning and decision making, as well as playing a role in life-and-death disasters. Thus, it is useful to explore the nature of cognitive biases and individual differences related to susceptibility or resistance to them. Salient issues include whether susceptibility to cognitive biases can be mitigated, such as through training and processes, like the structured analytic techniques employed by intelligence analysts (Heuer and Pherson, 2011), and the degree to which susceptibility is related to other human performance factors such as cognitive ability, working memory, executive functioning, and personality (see Chapter 2 for discussion of some of these factors).

THE NATURE AND DIVERSITY OF COGNITIVE BIASES

There have been several systematic attempts to catalog cognitive biases. To get a sense for the diversity of cognitive biases that have appeared in the literature, it is useful to note that Wikipedia[1] lists 92 "decision-making, belief, and behavioral biases," 27 "social biases," and 48 "memory errors and biases." Not all of these are distinct, and some may not be considered cognitive biases at all (i.e., the evidence for some of the listed biases, such as the bizarreness effect, is inconclusive), but it is a useful starting point.

---

[1] See "List of Cognitive Biases." Available: http://en.wikipedia.org/wiki/List_of_cognitive_biases [January 2015].

Cognitive Biases and Cognitive Ability

A question that arises is whether tasks that measure cognitive biases are measuring general cognitive ability. Stanovich and West (1998) investigated performance on tasks representing 28 cognitive biases (some based on prior literature and some first identified for the study) and found that for roughly half the tasks, there was no correlation with general cognitive ability. However, the denominator neglect problem is an example of a task for which there was a correlation: Participants are told they will win money by choosing a black marble in a tray of white and black marbles mixed as either 1 black in 10 marbles (10 percent chance of winning) or 8 black in 100 marbles (8 percent chance of winning). Participants tend to choose the 100-marble tray despite it being longer odds, perhaps with the idea that having 8 black marbles is interpreted as having 8 chances of winning, which is better than having only 1. However, participants with a higher general cognitive ability tended to choose the option with the better odds of winning, thereby demonstrating resistance to this type of cognitive bias. Another cognitive-ability-related task is the probabilistic reasoning task, in which respondents are asked to predict the number on the down side of 10 dealt cards when they are told that 7 cards have the number 1 and 3 cards have the number 2 on the down side (Stanovich and West, 1998). Most participants choose a strategy of predicting which 7 are 1 card and which 3 are 2 cards, even though a winning strategy is to predict 1-card status for all 10 cards. However, participants with higher cognitive ability are less likely to make this error.

An example of a task in which the cognitive bias is not correlated with cognitive ability is the anchoring effect task (Stanovich and West, 1998). In this task, participants are asked two questions, such as "Do you think there are more or less than 65 African countries in the United Nations?" and then "How many African countries do you think are in the United Nations?" Instead of "65" in the first question, half the participants were given the number "12." For those who were given "65" in the first question, the mean of their responses to the second question was 45.2; for those given "12" in the first question, the mean of the responses was 14.4. This discrepancy illustrates the anchoring effect, in that the information presented first heavily influenced the later decision. In this test, there was no correlation between SAT score and the size of the estimate in responding to the second question.

Another example of a task where the cognitive bias does not correlate with cognitive ability is the sunk cost task (Stanovich and West, 1998). Participants say that they would be more willing to drive an extra 10 minutes to save $10 on a $30 calculator than they would to save $10 on a $250 jacket, despite the fact that in either case the $10 savings is exactly the same.

The difference in willingness to drive an extra 10 minutes for the two items did not differ for groups with high versus low SAT scores. Another task for which there is no correlation with SAT scores is the "myside bias" task, related to confirmation bias, in which participants regardless of SAT scores are shown to more highly favor banning an unsafe German car in America than in having Germans ban an equally unsafe American car in Germany (Stanovich et al., 2013). Although biases are not necessarily detrimental (including, for example, "myside bias" may protect self-interests), the tendency toward them and their correlation with cognitive ability is important to understand in relation to performance.

An Individual-Differences Framework for Cognitive Biases

Oreg and Bayazi (2009) suggested that an individual-differences perspective could provide a theoretical framework for categorizing biases and could help account for the variance in predicting judgment and decision-making outcomes. Their framework suggests three categories of biases:

1. *Simplification biases* are motivated by comprehending reality, reflect information processes, and are related to cognitive ability and cognitive styles. Examples are denominator neglect—paying more attention to the number of times something has happened than to the number of opportunities for it to happen, such as believing that 1,286 cancer incidents out of 10,000 indicates a higher likelihood of cancer than 24.14 incidents out of 100 (Yamagishi, 1997)—and probability matching (for instance, if told that a card deck contains 60 percent red cards and 40 percent black cards, then when predicting the color of a card randomly drawn from the deck, the subject predicts "red" 60 percent of the time, rather than predicting "red" 100 percent of the time).
2. *Verification biases* are motivated by the desire to achieve consistency, reflect self-perception processes, and are related to core self-evaluation (which is a combination of self-efficacy [belief in one's ability to perform a task successfully] and locus of control [tendency to attribute successes and failures to one's own efforts and abilities rather than to external factors]). Examples are false consensus (believing others think like oneself) and learned helplessness (not acting due to prior experiences in which actions have not helped, even when actions would help in the current situation).
3. *Regulation biases* are motivated by the desire to approach pleasure and avoid pain, reflect decision-making processes, and are related to a person's approach/avoidance temperament. Examples are framing bias (being differentially sensitive to loss-and-gain fram-

ing) and endowment effects (once something is owned, its value increases). Also see Chapter 6 for a discussion of individual differences associated with abilities to function under circumstances of high emotion or "hot cognition" to include defensive reactivity, emotion regulation, and performance under stress.

Although this summary and framework are primarily rational, it seems that further research along these lines could validate or improve on this scheme and promote advances in understanding how cognitive biases can be integrated with other cognitive ability and personality factors research.

In addition to such trait factors being correlates of cognitive bias susceptibility, there may be other state factors that can affect decision making and susceptibility to cognitive biases. These include physical fatigue, sleeplessness, and emotional fatigue (or self-control depletion) (Muraven and Baumeister, 2000). For example, coping with stress, regulating negative affect, and resisting temptations have been found to affect subsequent self-control. The explanation has been that self-control is a limited resource, analogous to a muscle, and that continuous exercise of self-control degrades over time. If this is true then by exercising self-control, one might be more susceptible to inappropriate System 1 thinking, resulting in cognitive biases.

With respect to both trait and state cognitive bias factors, it seems reasonable that in an individual-differences framework their relationship could be fruitfully explored with potentially more powerful explanatory variables from an information processing perspective such as working memory, executive attention, and inhibitory control. These information-processing variables have been long known to correlate with general cognitive ability (e.g., Engle, 2002; Kyllonen and Christal, 1998), which is discussed in Chapter 2.

IARPA'S COGNITIVE BIAS MITIGATION PROGRAM

In 2011, the Intelligence Advanced Research Projects Activity (IARPA; a research entity within the Office of the Director of National Intelligence) announced the Sirius Program, whose goal was "to create experimental Serious Games to train participants and measure their proficiency in recognizing and mitigating the cognitive biases that commonly affect all types of intelligence analysis."[2] The program identified the following six cognitive biases for examination:

---

[2] Information is from the Sirius website, available: http://www.iarpa.gov/index.php/research-programs/sirius/baa. IARPA, Office of the Director of National Intelligence [January 2015].

1. Confirmation bias (interpreting events to support prior conclusions);
2. Fundamental attribution error (attributing events to others' personality rather than to circumstances);
3. Bias blind spot (not being aware of one's own biases);
4. Anchoring bias (overreliance on a single piece of information);
5. Representativeness bias (ignoring the base rate when categorizing or judging a likelihood of an event); and
6. Projection bias (attributing to others one's own beliefs, feelings, or values).

The significance of the IARPA project with respect to this report is twofold. First, the fact that bias mitigation strategies are being investigated in the Intelligence Community indicates the importance that community assigns to cognitive biases in judgment and decision making and to their broader significance and importance in intelligence analysis. Second, the identification of six specific cognitive biases suggests that these biases might be particularly important for intelligence analysts, a career field for civilian employees of the Intelligence Community as well as a military occupational specialty, and they may therefore warrant special attention in a program of research.

FUTURE RESEARCH

Challenging Issues

A number of important issues could be addressed in a broad program of research on cognitive biases. A key issue concerns individual differences and individual-level measurement: how can an individual's inherent susceptibility or resistance to cognitive biases best be measured? Much of the literature is concerned with documenting cognitive bias phenomena but is not concerned with developing individual measures of susceptibility to cognitive biases. A notable exception is the work on the cognitive reflection task (Toplak et al., 2014).

This distinction is important because many of the experimental designs in the cognitive bias literature operate differently depending on whether the bias manipulation is administered between or within groups. Consider, for example, the conjunction fallacy, which is the belief that it is more likely that someone is a member of both groups a and b than a member of just group a, after hearing a description that highlights group b traits. In the Linda Problem (Tversky and Kahneman, 1982; 1983), participants were told the following:

> Linda is 31-years-old, single, outspoken, and very bright. She majored in philosophy. As a student, she was deeply concerned with issues of discrimination and social justice, and also participated in antinuclear demonstrations.
>
> (Tversky and Kahneman, 1983, p. 297)

They then were asked "to check which of two alternatives was more probable":

> Linda is a bank teller.
> Linda is a bank teller and is active in the feminist movement.
>
> (Tversky and Kahneman, 1983, p. 299)

The finding was that 85 percent of the undergraduate respondents said that alternative 2 was more likely. According to Kahneman (2011) some studies did not show respondents both possibilities, as was done in the above example, but instead showed only one—either "bank teller" or "bank teller and is active in the feminist movement," that is a between-persons rather than within-persons design. In this case, in the between-persons design, the difference in the preference for the conjunction was even higher.

Tests of anchoring effects operate similarly: priming someone to guess high by presenting them with a high number is typically compared with priming *someone else* to guess low by presenting this different person with a low number. Low and high priming on the same person can be operationally difficult to test, due to carryover effects.

Much of cognitive bias research is based on a difference between conditions—one condition in which the bias is not invoked, and another in which it is. The challenges of using difference scores in within-person designs are a central part of cognitive bias measurement. In general, very little systematic research has been done on how best to measure the full range of cognitive biases.

Another issue concerns the theoretical structure of cognitive biases. It would be useful and desirable to have a general empirically based taxonomy of cognitive biases. The effort by Oreg and Bayazi (2009) seems to be a start, but considerable additional empirical work is required to develop such a taxonomy. Through such work, a more systematic taxonomy might lead, for instance, to a broader theoretical framework that enables predictions of an individual's susceptibility to biases, based on both individual and situational characteristics. It may also help answer questions about how cognitive biases are related. Is bias A simply a particular instance of bias B or a result of a very different set of mental processes? And even more important for military applications, can mitigation training on bias A result in reduction in susceptibility to both biases A and B?

Another key issue for theoretical explorations of cognitive biases has to do with their usefulness. Cognitive biases represent the application of thinking heuristics for problem solving; such heuristics are often useful shortcuts that enable faster decision making with less working memory burden. A good example is in medicine where physicians routinely use shortcuts or heuristics in their practice to sift through extensive information and formulate diagnoses (Vickrey et al., 2012). Heuristics are therefore not always ill-advised and do not always lead to improper decision making (Gigerenzer et al., 2011). A key research issue is when are they useful, and when should their use be curtailed? What are the training implications?

These research questions lead to a third key area for research, which concerns the effectiveness of training. If cognitive bias susceptibility is a relatively stable and enduring characteristic of individuals, a habitual way of thinking, then it might make sense at least for certain occupations to select out individuals with high susceptibility to cognitive biases. But if cognitive biases can relatively easily be effectively mitigated through training, then there may be less need to select for resistance to them. Currently far too little is known about the degree to which cognitive bias training is effective, how much is needed, and the degree to which training transfers to mitigation of related and of unrelated cognitive biases. Even if training is not effective, it could still be the case that system or job aids, such as the structured analytic techniques advocated within the Intelligence Community (Heuer and Pherson, 2011), can mitigate susceptibility to cognitive biases. For example, software might serve as a workaround to cognitive biases, but even less is known about this topic than about the preceding issues.

RESEARCH RECOMMENDATION

The U.S. Army Research Institute for the Behavioral and Social Sciences should support research to understand cognitive biases and heuristics, including but not limited to the following topics:

A. Research should be conducted to ascertain whether various cognitive biases and heuristics are accounted for by common bias susceptibility factors or whether various biases reflect distinct constructs (e.g., confirmation bias, fundamental attribution error).
B. A battery of cognitive bias and heuristics assessments should be developed for purposes of validity investigations.
C. Research should be conducted to examine the cognitive, personality, and experiential correlates of susceptibility to cognitive biases. This should include both traditional measures of personality and cognitive abilities (e.g., the Armed Services Vocational Aptitude Battery), and information-processing measures of factors such as working memory, executive attention, and inhibitory control.

D. Research should be conducted to identify contextual factors, that is, situations in which cognitive biases and heuristics may affect thought and action, and then to develop measures of performance in such situations, for use as criteria in studies aimed at understanding how cognitive biases affect performance. The research should consider the differentiating characteristics of contexts that determine when the use of heuristics for "fast and frugal" decision making might be beneficial, and when such thinking is better thought of as biased and resulting in poor decision making.

REFERENCES

Ariely, D. (2008). *Predictably Irrational: The Hidden Forces that Shape Our Decisions.* New York: Harper Collins.

Engle, R. (2002). Working memory capacity as executive attention. *Current Directions in Psychological Science,* 11(1):19–23.

Evans, J.St.B., and K.E. Stanovich. (2003). Dual-process theories of higher cognition: Advancing the debate. *Perspectives on Psychological Science,* 8(3):223–241.

Gigerenzer, G., R. Hertwig, and T. Pachur, Eds. (2011). *Heuristics: The Foundations of Adaptive Behavior.* New York: Oxford University Press.

Gilovich, T., and D.W. Griffin. (2002). Heuristics and biases: Then and now. In D.G.T. Gilovich and D. Kahneman, Eds., *Heuristics and Biases: The Psychology of Intuitive Judgment* (pp. 1–18). Cambridge, UK: Cambridge University Press.

Heuer, R.J., Jr. (1999). *The Psychology of Intelligence Analysis.* Center for the Study of Intelligence, Central Intelligence Agency. Available: https://www.cia.gov/library/center-for-the-study-of-intelligence/csi-publications/books-and-monographs/psychology-of-intelligence-analysis/ [January 2015].

Heuer R.J., Jr., and R.H. Pherson. (2011). *Structured Analytic Techniques for Intelligence Analysis.* Washington, DC: CQ Press.

Kahneman, D. (2011). *Thinking, Fast and Slow.* New York: Farrar, Straus, and Giroux.

Kyllonen, P.C., and R.E. Christal. (1990). Reasoning ability is (little more than) working-memory capacity?! *Intelligence,* 14(4):389–433.

Lewin, T. (2001). Sikh owner of gas station is fatally shot in rampage. *New York Times,* September 17, 2001. Available: http://www.nytimes.com/2001/09/17/us/sikh-owner-of-gas-station-is-fatally-shot-in-rampage.html [January 2015].

Muraven, M., and R.F. Baumeister, (2000). Self-regulation and depletion of limited resources: Does self-control resemble a muscle? *Psychological Bulletin,* 126(2):247–259.

Oreg, S., and M. Bayazi. (2009). Prone to bias: Development of a bias taxonomy from an individual differences perspective. *Review of General Psychology,* 13(3):175–193.

Roberto, M.A. (2002). Lessons from Everest: the interaction of cognitive bias, psychological, safety and system complexity. *California Management Review,* 45(1):136–158.

Ross, L. (1977). The intuitive psychologist and his shortcomings. In L. Berkowitz, Ed., *Advances in Experimental Social Psychology,* vol. 10. (pp.173–220). New York: Academic Press.

Spellman, B.A. (2011). Individual reasoning. In B. Fischhoff and C. Chauvin, Eds., *Intelligence Analysis: Behavioral and Social Scientific Foundations* (pp. 117–142). Washington, DC: The National Academies Press.

Stanovich, K.E., and R.F. West. (1998). Individual differences in rational thought. *Journal of Experimental Psychology: General, (127)*:161–188.

Stanovich, K.E., R.F. West, and M.E. Toplak. (2013). Myside bias, rational thinking, and intelligence. *Current Directions in Psychological Science, 22*(4):259–264.

Toplak, M.E., R.F. West, and K.E. Stanovich. (2014). Assessing miserly processing: An expansion of the Cognitive Reflection Test. *Thinking and Reasoning, 20*(2):147–168.

Tversky, A., and D. Kahneman. (1974). Judgment under uncertainty: Heuristics and biases. *Science, 185*(4,157):1,124–1,131.

Tversky, A., and D. Kahneman. (1982). Judgments of and by representativeness. In D. Kahneman, P. Slovic, and A. Tversky, Eds., *Judgment Under Uncertainty: Heuristics and Biases* (pp. 84–98). Cambridge, UK: Cambridge University Press.

Tversky, A., and D. Kahneman. (1983). Extension versus intuitive reasoning: The conjunction fallacy in probability judgment. *Psychological Review, 90*(4):293–315.

U.S. Department of Defense. (1988). *Investigation Report: Formal Investigation into the Circumstances Surrounding the Downing of Iran Air Flight 655 on 3 July 1988*. Washington, DC: U.S. Department of Defense. Available: http://www.dod.mil/pubs/foi/International_security_affairs/other/172.pdf [September 2014].

Vickrey, B.G., M.A. Samuels, and A.H. Ropper. (2010). How neurologists think: A cognitive psychology perspective on missed diagnoses. *Annals of Neurology, 67*(4):425–433.

Yamagishi, K. (1997). When a 12.86% mortality is more dangerous than 24.14%: Implications for risk communication. *Applied Cognitive Psychology, 11*(6):495–506.

4

Spatial Abilities

Committee Conclusion: A spatial ability measure, Assembling Objects (AO), is included in the Armed Services Vocational Aptitude Battery (ASVAB). Research suggests incremental validity for spatial measures over general mental ability measures in predicting important military outcomes. Research also suggests that sex differences vary across different operationalizations of spatial ability. Together, these findings suggest exploring varying approaches to the measurement of spatial abilities to ascertain whether the AO test is the best measure of spatial ability for military selection and classification. The committee concludes that spatial ability merits inclusion in a program of basic research with the long-term goal of improving the Army's enlisted accession system.

The current ASVAB is largely a measure of acquired knowledge and ability (see Roberts et al., 2000). The potential for developing measures of fluid intelligence as a supplement to the current ASVAB is treated in detail in Chapter 2. Another domain in which skill is generally not acquired by formal instruction is that of spatial ability: the capacity to unravel, understand, and remember the spatial relations among objects. The AO subtest of the ASVAB is an indicator of this skill, but is not currently used in Army selection or placement decisions. Spatial ability is not a monolithic and static trait, but made up of numerous subskills, which are interrelated among each other and develop throughout a lifetime. While the committee treats this topic separately in this chapter, it is important to examine how individual differences in spatial abilities intersect with, and are distinct from, the cognitive-control and inhibitory capabilities covered in Chapter 2, the cognitive biases covered in Chapter 3, and the "hot cognition" processes covered in Chapter 6.

An argument can be put forward that this kind of visual-spatial ability is becoming increasingly important with the development and proliferation of new technologies, such as imaging, computer graphics, data visualization, and supercomputing. Highly demanding spatial tasks include the construction of mental representations of object configuration from images on several screens representing different perspectives, as in some fields of interest to the military. In these fields of work, powerful computer graphic technologies are being used to create complex visual images of processes that occur in the natural world. Despite their importance in many fields and in science education, spatial skills rarely work in isolation from other abilities, such as logical reasoning, efficient memory retrieval, and verbal skills, and deficits in one area can often be compensated for by excellence in others. An important type of exceptional talent in math and science, however, is the ability to easily switch from one efficient mode of representation to another (e.g., from a conceptual to a spatial mode and vice versa).

It is clear that spatial abilities can be measured in a large-scale group setting (as is done with the current ASVAB) and contribute to military performances (e.g., relationship with hands-on performance tasks; see Carey, 1994). The individual-differences literature is replete with recent articles describing the importance of rapid stimulus selection and thinking with symbols (examples are Hegarty and Waller, 2005; Lathan and Tracey, 2002; Malinowski and Gillespie, 2001). Spatial abilities have been found to be predictive of real-life events (Carey, 1994), including map reading and arterial positioning (see McHenry et al., 1990, for findings from Project A data).

DEFINING SPATIAL ABILITIES

A recent upsurge of empirical evidence suggests spatial abilities are an important predictor of performance (see Lubinski, 2010), especially in scientific and technical fields (National Research Council, 2006; Shea et al., 2001; Stieff et al., 2014; Wai et al., 2009). Spatial abilities are important for understanding an individual's spatial relationship to and within surroundings (e.g., orienteering) and also for understanding representations of multidimensional figures in one-dimensional displays (e.g., data visualization). Within visual perception abilities, spatial abilities can be defined as "how individuals deal with materials presented in space—whether in one, two, or three dimensions, or with how individuals orient themselves in space" (Carroll, 1993, p. 304). Furthermore, spatial abilities signify "an ability in manipulating visual patterns, as indicated by level of difficulty and complexity in visual stimulus material that can be handled successfully, without regard to the speed of task solution" (Carroll, 1993, p. 362).

Spatial abilities are multifaceted, and tests to measure individual differences in spatial abilities must separate these facets to distinguish them within the domain of spatial abilities as well as from other measures of general intelligence. Spatial ability tests measure practical and mechanical abilities important for success in technical occupations, but they are not supposed to be measures of abstract reasoning abilities (Horn, 1989; Smith, 1964). Similarly, test design is challenged by the important role afforded to spatial imagery in accounts of creative thinking (Shepard, 1978) and for the observed high and positive correlations between spatial ability tests and other measures of intelligence (for reviews, see Lohman, 1996; Lohman et al., 1987). Furthermore, spatial abilities measure psychological factors such as attention, important for everyday demands on working memory to maintain and transform images (Kyllonen and Christal, 1990).

The utility of spatial abilities as performance predictors has a long history of research, test development, and longitudinal outcome assessment (with some results being contradictory and debated). Many researchers have expressed the view that spatial abilities are as important as verbal comprehension in the prediction of real life events (for overview, see Humphreys and Lubinski, 1996). It is difficult to distinguish spatial abilities from fluid intelligence or broad reasoning (Horn, 1989) mainly because most of the test material for fluid intelligence is visual in nature. Examples include Project Talent's Abstract Reasoning (Project Talent Office, 1961) and Raven's Progressive Matrices (Raven, 1992). Research also shows that prevalent standardized tests of cognitive abilities fail to identify talent for outstanding achievement in domains not conducive to recognition or expression through verbal mechanisms prevalent in modern academic and testing realms (Lohman, 1994; 2005), a finding demonstrated even among those in the highest tiers of general cognitive ability (Kell et al., 2013; Robertson et al., 2010).

TESTING SPATIAL ABILITIES FOR MILITARY ENTRANCE

The military's interest in spatial abilities testing dates back to World War I, and by WWII a spatial-visualization test was included in the Army General Classification Test (for a historical overview, see Humphreys and Lubinski, 1996). The range of military occupational specialties to which the military services select and assign recruits, including many that demand abilities beyond verbal and mathematical reasoning, suggests that inclusion of measures of spatial ability in the military's entrance test battery would facilitate the identification of potentially highly successful recruits who might otherwise be overlooked or placed in suboptimal occupations.

The value of spatial abilities tests is well known to the U.S. Army Research Institute for the Behavioral and Social Sciences (ARI). The current ASVAB includes AO, a test for a specific facet of spatial abilities (Powers, 2013). The Army's Selection and Classification Project (Project A), conducted by ARI, identified AO as a potential performance predictor in Army occupations (Buscigilo et al., 1994; Campbell and Knapp, 2001). Currently, AO is assessed through 25 questions (tested in 15 minutes) on the paper-and-pencil ASVAB version and through 16 questions (tested in 16 minutes) in the computer adaptive version.[1] AO tests the individual's "ability to determine correct spatial forms from separate parts and connection points" (Held and Carretta, 2013, p. 2). Figure 4-1 displays two sample questions from the AO test.

AO was included along with several other spatial abilities tests as part of the Enhanced Computer-Administered Test (ECAT) battery described by Alderton and colleagues (1997). Wolfe (1997) reported incremental validity of .013 for a composite ECAT spatial score over the current ASVAB for predicting training school grades, and incremental validity of .03 for performance on hands-on performance tests. Additionally, Carey (1994) reported that AO added incremental validity to the ASVAB in predicting mechanics' job performance in a hands-on performance test of both automotive (.012) and helicopter mechanics (0.15). Thus the incremental validity of AO is modest and varies by criteria. But modest increments can be of considerable applied utility in settings where large numbers of screening decisions are made.

At this time, AO scores are not used for selection purposes by any military service (the scores are not part of the Armed Forces Qualification Test, AFQT), and only the Navy currently uses the AO results for occupational classification (Held and Carretta, 2013).

Consistent with Lohman's (1979) identification of three basic spatial abilities factors, analysis of spatial abilities in ARI's Project A considered spatial relations, spatial orientation, and visualization. The currently used AO test was developed through job analysis of abilities important to Army occupations during Project A's study time frame, 1983 to 1988, and refined through subsequent field testing and validation studies. As discussed below, some tests of spatial abilities demonstrate subgroup differences between sexes; however, such differences were not found in AO (Peterson et al., 1990). For this reason in particular, the Navy adopted AO for classification purposes, as it sought to reduce or eliminate adverse impact for minority groups seeking technical occupations. Sex and minority-group differences in performance on AO have been shown to be lower than for some other

---

[1] See http://www.official-asvab.com/whattoexpect_rec.htm [June 2014] for the most up-to-date information on the content of the ASVAB.

Assembling Objects (AO)

Question 1. Which figure best shows how the objects in the left box will touch if the letters for each object are matched?

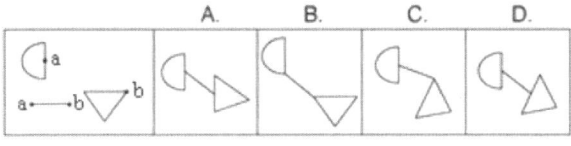

○ A.
○ B.
○ C.
○ D.

Assembling Objects (AO)

Question 2. Which figure best shows how the objects in the left box will appear if they are fit together?

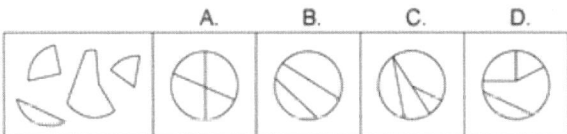

○ A.
○ B.
○ C.
○ D.

FIGURE 4-1 Sample questions from the Armed Services Vocational Aptitude Battery AO test.
SOURCE: Official site of the ASVAB. Assembling Objects. Available: http://official-asvab.com/questions/app/question_ao1_app.htm [January 2015].

technical tests, such as Auto and Shop Information, which has the largest effect size among the ASVAB subtests (Held and Carretta, 2013). Furthermore, a recent ARI assessment of AO (Anderson et al., 2011) concluded that adding AO to the AFQT composite score would increase the AFQT's prediction of performance and job knowledge in jobs that require spatial aptitude, with little or no subgroup differences.

STATE OF THE SCIENCE

As a leader in research on spatial abilities, the Army has access to a well-developed body of data on the internal and external validity of spatial ability constructs and tests. However, in considering the historical timeline of much of this research, the increased roles of technology in many modern military occupations, and developments in testing methods, the committee finds the science behind the predictive power of spatial abilities worthy of further exploration and consideration. Much has been learned about spatial abilities since the research findings conducted through Project A that developed the AO test, and modern technological developments have implications for many of the duties and tasks essential to soldier performance, as well as implications for the Army's ability to test for those abilities. While much of the research conducted on spatial abilities has been and will continue to be conducted within applied research programs, there remains a great deal of foundational knowledge that is still needed about the multiple facets of spatial abilities. The following section briefly reviews the current state of the science in understanding and testing for spatial abilities.

Internal Validity: Measuring Spatial Abilities

The multifaceted nature of spatial abilities poses a challenge to identify and develop measures of the separate facets, as well as to understand their interrelationships. Since the early 1900s, researchers have sought evidence of a general spatial factor and its testable component parts. In addition to Lohman's three basic factors (spatial relations, spatial orientation, and visualization), other major spatial factors identified and assessed over a century of research include, for example (see Carroll, 1993, for an overview):

- *Visual memory*—short-term memory of visually presented stimuli;
- *Spatial scanning*—tasks, such as following a maze or selecting a path;
- *Perceptual speed*—rate at which presented visual stimuli are matched;
- *Serial integration*—identification of pictures presented successively;

- *Closure speed*—time to match an incomplete visual presentation to a known object or feature: and
- *Kinesthetic*—coordinated body motion such as left-right judgments.

In a recent meta-analysis of studies on the trainability of spatial skills, Uttal and colleagues (2013) noted a lack of consensus on a typology of spatial abilities and proposed replacing the "spatial relations, spatial orientation, and visualization" typology with a classification system with two fundamental distinctions: intrinsic versus extrinsic information and static versus dynamic tasks. The authors found their typology more useful in categorizing studies for their meta-analysis of trainability studies than the typology based on factor-analytic studies.

Most research on spatial abilities has focused on what may be termed "small-scale" tasks (e.g., mental rotation of objects), with a focus on "large-scale" tasks (e.g., navigation, way-finding) emerging more recently (Hegarty et al., 2006). Recent work, culminating in a meta-analysis by Wang and colleagues (2014), argues that these are best viewed as two separate families of abilities. However, the mean correlation of .27 found in this meta-analysis between the two families indicates that they are not completely independent. Additional evidence of their separability comes from the finding that different areas of the brain have been identified as involved in these two families of abilities, with small-scale tasks linked to activation of the parietal lobes and large-scale tasks linked to the hippocampus and medial lobes (Kosslyn and Thompson, 2003; Morris and Parslow, 2004).

This distinction between small-scale and large-scale spatial abilities links to recent U. S. Air Force research on the construct of *situation awareness*. As defined by Dr. Mica Endsley, U.S. Air Force Chief Scientist, in her briefing to the committee, "Situation awareness is the perception of elements in the environment within a volume of time and space, the comprehension of their meaning, and the projections of their status in the near future."[2] The inclusion of a future time element in this definition of situation awareness indicates the importance of abilities in contingency planning, a partially trainable skill with underlying individual differences in natural ability. The committee can envision facets of spatial abilities similarly linked with situation awareness also being important to many Army occupations and job duties, especially those involving combat maneuvers of remotely controlled technology.

To test individuals' situation awareness abilities, Dr. Endsley described a direct measure, the Situation Awareness Global Assessment Technique,

---

[2] Presentation to the committee on December 5, 2013. Presentation cited work contained in Endsley (1988). Full presentation materials available by request through this study's public access file.

which has been validated for content (Endsley, 1993), construct (Endsley, 2000), and criterion (Endsley, 1990). This assessment technique requires subjects to monitor a simulation that is randomly frozen with all displays blanked out. The subject must then answer a series of rapid questions about the state of the simulation to assess the subject's situation awareness at a specific point in time, including the subject's assessment of the expected situation in the near future. In assessing situation awareness, an important distinction must be made between decision making and situation awareness. An individual's ability to identify and comprehend the significance of available information about his or her environment is an important factor of spatial abilities that impacts the individual's ability to make effective and accurate decisions. Although not always the case, poor decisions, sometimes with fatal consequences, have resulted from poor situation awareness whereby the decision maker fails to recognize important details about the environment (also see Chapter 3 for a discussion of the role cognitive biases play in the interpretation of available information). For a more detailed discussion of the critical importance of situational understanding for decision making in an Army context, see the National Research Council report, *Making the Soldier Decisive on Future Battlefields* (National Research Council, 2013).

External Validity: Using Spatial Abilities in Predictions of Key Outcomes

Much of the work on spatial abilities by Project A focused on the demands of Army occupations. Campbell and Knapp (2001) documented the incremental validity of AO during Project A for predicting a number of criteria important to the Army; Alderton and colleagues (1997) and Wolfe (1997) described the absolute and incremental validities of several spatial abilities measures across all military services. More recently, Anderson and colleagues (2011) concluded that adding AO to the composite of the ASVAB that is used for selection into the Army would increase the validity of that composite for many criteria.

Other evidence of the validity of constructs for facets of spatial ability in predicting key outcomes was obtained through a longitudinal study of the occupational status of 400,000 high school students from Project Talent (see Humphreys et al., 1993). Using self-report interest questionnaires and ability tests, Austin and Hanisch (1990) predicted occupational groups for a mixed-gender sample of 10th graders included in Project Talent. Prediction accuracy varied across groups "as a function of the a priori selection of the specific occupational groups that formed the criterion categories" (Humphreys et al., 1993, p. 251). Austin and Hanish (1990) extracted and interpreted five discriminant functions; the first function was dominated by verbal and mathematical tests, while the second function was dominated by mechanical, spatial, and mathematical tests. These first two functions accounted for the major proportion of the variance, while the other three functions, which included dimensions of various vocational interests, accounted for small but significant proportions of the variance.

Humphreys and colleagues (1993) were among the influential researchers on spatial abilities. The spatial composite they used was made up of four tests: (1) the Project Talent 2D spatial abilities test (object rotation and flipping in two dimensions; 24 items); (2) the Project Talent 3D spatial abilities test described above (three-dimensional test of mental folding; 16 items); (3) the Project Talent Mechanical Reasoning test (20 items), which measured deductions based on primitive mechanisms (e.g., gears, pulleys, and springs) and knowledge of the effects of common physical forces (e.g., gravity); and (4) Abstract Reasoning (15 items), which was a nonverbal test of logical relationships in complex figural patterns. The first two tests are relatively pure measures of spatial abilities, whereas the last two are likely to be visual measurements of a broad reasoning (fluid intelligence) skill.

Humphreys and colleagues (1993) reported on what they termed High-Space students (the highest 20 percent of scorers on a spatial-mathematics composite, resulting in an N of 17,647). This group was distinguished from the group they termed High-Intelligence students (the highest 20 percent across both a verbal-mathematics composite and the spatial-mathematics composite, resulting in an N of 54,311). Both males and females in the High-Space group avoided educational and occupational opportunities, relatively speaking, in the social sciences and humanities over the course of 11 years following high school graduation. Both genders in the High-Space group, but especially the males, were working in larger proportion in traditional blue-collar occupations. See Table 4-1 for data that demonstrates a commonality between engineers, artists, and artisans.

The High-Space group also had substantially fewer completed degrees at every educational level beyond high school graduation, compared with the High-Intelligence and High-Verbal groups (see Table 4-2). The authors reported that the predictive validities of the spatial-mathematics and verbal-mathematics ability composites were established by successfully differentiating a variety of educational and occupational groups. By relying solely on scores from conventional mathematical and verbal ability tests, such as those of the Scholastic Aptitude Test and the Graduate Record Examination, physical science and engineering disciplines may be failing to identify and select highly talented individuals.

With respect to the value of a spatial-mathematics composite measure, the authors concluded

TABLE 4-1 Proportions in Four High School Classes of Occupational Categories of Three Select Groups and Students in General

Major	Norm		High-Intelligence		High-Space		High-Verbal	
	Males	Females	Males	Females	Males	Females	Males	Females
Physical Science	.04	.00	.15	.01	.10	.00	.04	.00
Biological Science	.02	.03	.07	.05	.04	.06	.06	.07
Business	.32	.07	.34	.08	.35	.06	.34	.07
Education/Social Work	.07	.08	.09	.18	.09	.10	.14	.19
Humanities/Social Sciences	.04	.01	.11	.04	.02	.01	.19	.04
Arts	.01	.01	.01	.01	.03	.01	.02	.00
Technical	.06	.02	.08	.02	.08	.02	.05	.02
Secretarial/Clerical	.04	.16	.02	.11	.02	.16	.01	.12
Artisan	.39	.11	.14	.05	.28	.07	.12	.05
Housewife	.00	.50	.00	.44	.00	.50	.00	.43

NOTE: These proportions are based on population estimates.
SOURCE: Humphreys et al. (1993, p. 256).

TABLE 4-2 Proportions in Four High School Classes of Amount of Education Completed by the Three Select Groups and Students in General

Education	Norm		High-Intelligence		High-Space		High-Verbal	
	Males	Females	Males	Females	Males	Females	Males	Females
PhD	.03	.00	.11	.02	.02	.00	.08	.01
MA+	.06	.04	.16	.13	.09	.05	.18	.09
BA+	.17	.13	.34	.33	.25	.17	.34	.38
HS+	.53	.64	.26	.41	.50	.65	.27	.41
Dropped Out	.05	.06	.01	.01	.01	.04	.01	.01
Uncertain	.02	.01	.03	.02	.01	.01	.02	.02
No Response	.14	.12	.10	.08	.11	.09	.10	.08

NOTE: These proportions are based on population estimates. PhD = doctoral degree; it includes degrees in law and medicine. The + next to MA (master's degree in arts), BA (baccalaureate degree in arts), and HS (high school diploma) = the inclusion of individuals having course work beyond that level but not enough to achieve the next highest credential.
SOURCE: Humphreys et al. (1993, p. 257).

Scores on a spatial-visualization composite would probably add incremental validity to verbal and math scores, which are currently being used for identifying students with exceptional talent for engineering and physical science. Moreover, spatially talented individuals not only have the ability to achieve career excellence in engineering and the physical sciences but they also are more likely to remain committed to these disciplines. Furthermore, although our research was aimed at the more technical sciences, we found that the importance of spatial skills is also seen in many of the creative arts. . . . The prevailing emphasis on verbal scores on national tests and on grades in verbal courses for placing students in the precollege curriculum and in encouraging students to think of themselves as college material might be destructive to those who are intellectually talented in nonverbal ways. Students who are fluent verbally are ideal in the minds of many educational personnel at all levels, and this ideal is readily transmitted to parents and students. The case must be made for another important combination of abilities, and students who are suitably high on that combination should be strongly encouraged to aspire to college training. Consequently, more spatially talented students could be entering technical disciplines (which are highly correspondent to their abilities and interests).
(Humphreys et al., 1993, pp. 258–259)

Additional research shows that spatially talented youth are an important pool of human capital in science, technology, engineering, and mathematics fields, and "the influence of this intellectual pattern extends beyond learning and work settings and into domains of creative production" (Kell et al., 2013, p. 1,835). Furthermore, the correlation between spatial ability and career choice, and performance and persistence in that career, is demonstrated even in the top one percent of adolescents in cognitive ability (Robertson et al., 2010). This suggests that even those recruits scoring in the highest tiers of the AFQT would benefit from job assignments made in consideration of their talents in spatial abilities.

Sex Differences

The adoption of tests of spatial abilities for the purposes of assessment and placement has been hampered by contradictory and sometimes confusing results regarding sex differences. Males score higher, on average, than females in many spatial abilities tests, but this is not true across all facets of spatial ability. In fact, the current AO test in the ASVAB reduces adverse impact and score barriers for both women and ethnic/racial minority groups (Held and Carretta, 2013). Other research indicates that men and women do not, as subgroups, show the same score distributions on many tests of facets of spatial ability (Jones and Anuza, 1982), and long-standing questions remain in dispute about the magnitude of the differences, as well as the facets of spatial ability in which sex differences in score distribution are apparent and the age at which those differences appear (Linn and Petersen, 1985).

Humphreys and Yao (2002) used Project Talent data from 57 cognitive tests to analyze college-major selection preferences. Based on what they termed their Descriptive Discriminant Analysis, they concluded, "Large sex differences in the incidence of various choices do not affect appreciably patterns of scores on cognitive and self-report tests" (Humphreys and Yao, 2002, p. 8). While it appears that men score higher then women on this kind of test, Humphreys and Yao found that, for both sexes, science majors are identifiable from those in the humanities and social sciences through a combination of mechanical, spatial, and mathematical tests.

Voyer and colleagues (1995) conducted a meta-analysis of 286 effect sizes from studies employing a variety of spatial ability measures. They found significant sex differences in several tests, and they found that differences from test to test still exist. Military research also shows variability in sex differences across spatial ability measures. Russell and Peterson (2001) reported a male-female d = 0.06 for AO, but d values greater than .30 for three other spatial tests.[3] Voyer and colleagues (1995) also found some support for the belief that sex differences on spatial ability measures have decreased recently. Finally, it was found that the age of emergence of sex differences varied by the type of test administered. The data from this meta-analysis leads the committee to believe further research is necessary to better understand sex differences in spatial ability.

Interestingly, the relationship between spatial ability and verbal ability has been found in some studies to differ between men and women. For example, in a test of fluid intelligence using letters (Primary Mental Abilities Battery), females outperformed males. However, in a similar test using figures rather than letters (Advanced Progressive Matrices Test), males outperformed females (Colom and García-López, 2002).

The debate surrounding possible differences in experience and processing of spatial information between men and women has been heightened by research on the impact of video games on spatial abilities test scores. Feng and colleagues (2007, p. 850) noted that "boys have always played different games than girls, and early recreational activities have often been cited as a major cause of gender differences in adult spatial cognition." While it is important to note that spatial abilities do appear to be amenable to training (Feng et al., 2007; we note that this study makes use of a very small sample size and should be viewed as merely suggestive), a key question is whether training on one task generalizes to other tasks. Sims and Mayer (2002)

---

[3] "d is the standardized mean difference between group means" (Russell and Peterson, 2001, p. 275).

found that experience with the computer game Tetris aided performance on a Tetris-like task, but not on other spatial tasks, leading them to conclude that spatial training effects were task-specific. Terlecki and colleagues (2008) also found evidence of transfer from Tetris training to performance on mental rotation tasks. The committee's interpretation of this body of research is that spatial ability training is usually domain specific and does not generalize to other dissimilar domains (e.g., training spatial ability in chemistry students for detecting the chirality of molecules does not increase ability to identify and recall the spatial locations of truck engine parts).

Regarding training effects on gender differences, the classic meta-analysis by Baenninger and Newcombe (1989) reported that training produces comparable changes in performance for males and females, and hence produces no reduction in the mean difference. A number of more recent studies have reported differing findings, with gender differences reduced or even eliminated. For example, Stieff and colleagues (2014) focused on strategy training in problem solving and reported that specific training on mental imagery with additional training in analytic strategies eliminated sex differences in achievement. However, a meta-analysis by Uttal and colleagues (2013, p. 367) of the trainability of spatial abilities concluded that, "Both men and women responded substantially to training; however, the gender gap in spatial skills did not shrink due to training." The committee gives these meta-analytic findings much greater weight than the findings of small-scale individual studies. Given the varieties of spatial abilities and the range of possible training interventions, we view the issues as not yet fully resolved.

The focus of this section on sex differences in spatial abilities reflects the considerable attention such differences have received in the recent research literature. In particular, the committee's interest in sex differences is based on findings that male-female differences on some spatial ability measures are substantially greater, in proportional terms, than are sex differences on the currently used AFQT composite. For instance, Sackett and colleagues (2009) report a male-female AFQT d of 0.08 for a nationally representative sample of 18-22 year old youth, which is similar to the 0.06 reported in the Russell and Peterson Project A data for AO, but quite different from the d's greater than 0.30 that Russell and Peterson (2001) reported for three other spatial tests (maze, orientation, and map). However, the committee notes that racial/ethnic mean differences are also generally found on spatial ability measures. For example, Russell and Peterson (2001) reported white-black d's ranging from 0.69 to 1.08 for six spatial ability measures in Army Project A data. These are substantially larger differences than those Russell and Peterson reported for male-female differences (see above). The focus above on sex differences does not imply that racial/ethnic group differences are not present or are not important for the Army's accession and selection purposes.

New Technologies

The rate of advance in modern technologies has implications for two major aspects of the Army's selection and assignment process: (1) the duties and tasks of many technical fields for which recruits will be assigned, and (2) the available methods of testing recruits for spatial abilities. The Army's primary work on spatial abilities was conducted through Project A, which assessed the spatial abilities relevant to Army occupations at that time. Modifications of jobs and occupations, especially during the past 30 years, warrant a new look at the fundamental importance of spatial abilities to Army occupations, including those aspects of spatial ability that require proficiency in human-computer systems, virtual interfaces, graphical data representations, and other digital-age technologies. The intrinsic capabilities of many of these everyday work duty technologies also provide opportunities for new methods of assessing spatial ability (see Section 5, Methods and Methodology, for further discussion). With the decrease of pencil-and-paper ASVAB testing in favor of the computer adaptive CAT-ASVAB, it is now feasible to consider test formats that were previously not easily implemented, when only the paper-and-pencil format was available. More important, inevitable advances in hardware and software capabilities will make it possible to consider tests dependent upon such three-dimensional features as precise measurement of speed, introduction of eye-tracking visual search, and real-time three-dimensional spatial manipulation of virtual objects.

A PATH FORWARD

Research indicates that spatial abilities are distinct from general intelligence. Furthermore, these abilities are separable and specific; some facets of spatial ability can be measured now, while others should be measurable in the near future. Nonetheless, there are numerous open questions about spatial abilities. Uttal and colleagues (2013, p. 370) observe, " . . . much of the focus of research on spatial cognition and its development has been on the biological underpinnings of these skills. . . . Perhaps as a result, relatively little research has focused on the environmental factors that influence spatial thinking and its improvement." As noted by a previous National Research Council committee, "Through the support of federal funding agencies . . . there should be a systematic research program into the nature, characteristics, and operations of spatial thinking" (National Research Council, 2006, p. 7).

Ultimately, the Army would benefit from learning the utility of spatial abilities in determining the (a) initial selection of recruits, (b) their preferred choices for occupations, (c) their actual classification into occupations, (d) their long-term retention in those occupations (and in the Army), and (e) their performance in those occupations. The path to addressing those important questions first requires answers to more basic questions about spatial abilities in general, how they can be developed (or trained), and how they can be most appropriately measured.

RESEARCH RECOMMENDATION

The U.S. Army Research Institute for the Behavioral and Social Sciences should support research to understand facets and assessment methods in the domain of spatial abilities, including the following research lines of inquiry:

A. Identify or develop measures of various facets of spatial ability, with particular attention to the role of technology to overcome prior limitations in test-item formats.
B. Examine the interrelationships among various facets of spatial ability, including but not limited to spatial relations, spatial orientation, and spatial visualization.
C. Examine sex differences on the various facets of spatial ability, as well as the degree to which sex differences are mitigated or accentuated by various forms of training on the facets of spatial ability.
D. Develop measures reflecting various work outcomes that can be used as criterion measures in evaluating the validity of various measures of spatial ability.

REFERENCES

Alderton, D.L., J.H. Wolfe, and G.E. Larson. (1997). The ECAT battery. *Military Psychology,* 9(4):5–38.
Anderson, L., R.R. Hoffman, B. Tate, J. Jenkins, C. Parish, A. Stachowski, and J.D. Dressel. (2011). *Assessment of Assembling Objects (AO) for Improving Predictive Performance of the Armed Forces Qualification Test.* Arlington, VA: U.S. Army Research Institute for the Behavioral and Social Sciences.
Austin, J.T., and K.A. Hanisch. (1990). Occupational attainment as a function of abilities and interests: A longitudinal analysis using Project Talent data. *The Journal of Applied Psychology,* 75(1):77–86.
Baenninger, M., and N. Newcombe. (1989). The role of experience in spatial test performance: A meta-analysis. *Sex Roles,* 20(5-6):327–344.
Buscigilo, H.H., D.R. Palmer, I.H. King, and C.B. Walker. (1994). *Creation of New Items and Forms for the Project A Assembling Objects Test.* Technical Report 1004. Alexandria, VA: U.S. Army Research Institute for the Behavioral and Social Sciences.

Campbell, J.P., and D.J. Knapp, Eds. (2001). *Exploring the Limits in Personnel Selection and Classification*. Mahwah, NJ: Lawrence Erlbaum Associates.
Carey, N.B. (1994). Computer predictors of mechanical job performance: Marine Corps findings. *Military Psychology*, 6(1):1–30.
Carroll, J.B. (1993). *Human Cognitive Abilities: A Survey of Factor-Analytic Studies*. Cambridge, UK: University of Cambridge.
Colom, R., and O. García-López. (2002). Sex differences in fluid intelligence among high-school graduates. *Personality and Individual Differences*, 32(3):445–451.
Endsley, M.R. (1988). Design and evaluation for situation awareness enhancement. *Proceedings of the Human Factors Society 32nd Annual Meeting*, 32(97):97–101. Available: http://pro.sagepub.com/content/32/2/97.abstract [September 2014].
Endsley, M.R. (1990). A methodology for the objective measurement of situation awareness. In Advisory Group for Aerospace Research and Development, Conference Proceedings No. 478, *Situational Awareness in Aerospace Operations* (pp. 1/1-1/9). Neuilly Sur Seine, France: NATO–AGARD. Available: http://ftp.rta.nato.int/public//PubFullText/AGARD/CP/AGARD-CP-478///AGARD-CP-478.pdf [January 2015].
Endsley, M.R. (1993). Situation awareness and workload: Flip sides of the same coin. In R.S. Jensen and D. Neumeister, Eds., *Proceedings of the Seventh International Symposium on Aviation Psychology* (pp. 906–911). Columbus: Ohio State University.
Endsley, M.R. (2000). Direct measurement of situation awareness: Validity and use of SAGAT. In M.R. Endsley and D.J. Garland, Eds., *Situation Awareness Analysis and Measurement* (pp. 147–174). Hoboken, NJ: Taylor and Francis.
Feng, J., I. Spence, and J. Pratt. (2007). Playing an action video game reduces gender differences in spatial cognition. *Psychological Science*, 18(10):850–855.
Hegarty, M., and D. Waller. (2005). Individual differences in spatial abilities. In P. Shah and A. Miyake, Eds., *The Cambridge Handbook of Visuospatial Thinking* (pp. 121–169). New York: Cambridge University Press.
Hegarty, M., D.R. Montello, A.E. Richardson, T. Ishikawa, and K. Lovelace. (2006). Spatial abilities at different scales: Individual differences in aptitude-test performance and spatial-layout learning. *Intelligence*, 34(2):151–176.
Held, J.D., and T.R. Carretta. (2013). *Evaluation of Tests of Processing Speed, Spatial Ability, and Working Memory for Use in Military Occupational Classification* (NPRST-TR-14-1). Millington, TN: Navy Personnel Research, Studies, and Technology.
Horn, J.L. (1989). Models for intelligence. In R. Linn, Ed., *Intelligence: Measurement, Theory, and Public Policy* (pp. 29–73). Urbana: University of Illinois Press.
Humphreys, L.G., and D. Lubinski. (1996). Assessing spatial visualization: An underappreciated ability for many school and work settings. In C.P. Benbow and D. Lubinski, Eds., *Intellectual Talent: Psychometric and Social Issues* (pp. 116–140). Baltimore, MD: Johns Hopkins University Press.
Humphreys, L.G., and G. Yao. (2002). Prediction of graduate major from cognitive and self-report test scores obtained during the high school years. *Psychological Reports*, 90(1):3–30.
Humphreys, L.G., D. Lubinski, and G. Yao. (1993). Utility of predicting group membership and the role of spatial visualization in becoming an engineer, physical scientist, or artist. *Journal of Applied Psychology*, 78(2):250–261.
Jones, B., and T. Anuza. (1982). Effects of sex, handedness, stimulus and visual field on "mental rotation." *Cortex*, 18(4):501–514.
Kell, H.J., D. Lubinski, C.P.O. Benbow, and J.H. Steiger. (2013). Creativity and technical innovation: Spatial ability's unique role. *Psychological Science*, 24(9):1,831–1,836.
Kosslyn, S.M., and W.L. Thompson. (2003). When is early visual cortex activated during visual mental imagery? *Psychological Bulletin*, 129(5):723–746.

Kyllonen, P.C., and R.E. Christal. (1990). Reasoning ability is (little more than) working memory capacity?! *Intelligence, 14*(4):389–433.

Lathan, C.E., and M. Tracey. (2002). The effects of operator spatial perception and sensory feedback on human-robot teleoperation performance. *Presence, 11*:368–377.

Linn, M.C., and A.C. Petersen. (1985). Emergence and characterization of sex differences in spatial ability: A meta-analysis. *Child Development, 56*(6):1,479–1,498.

Lohman, D.F. (1979). *Spatial Ability: A Review and Re-analysis of the Correlational Literature* (Technical Report No. 8.). Stanford, CA: Aptitudes Research Project, School of Education, Stanford University.

Lohman, D.F. (1994). Spatially gifted, verbally inconvenienced. In N. Colangelo, S.G. Assouline, and D.L. Ambroson, Eds., *Talent Development: Vol. 2. Proceedings from the 1993 Henry B. and Jocelyn Wallace National Research Symposium on Talent Development* (pp. 251–264). Dayton: Ohio Psychology Press.

Lohman, D.F. (1996). Spatial ability and G. In I. Dennis and P. Tapsfield, Eds., *Human Abilities: Their Nature and Assessment* (pp. 97–116). Hillsdale, NJ: Lawrence Erlbaum Associates.

Lohman, D.F. (2005). The role of non-verbal ability tests in identifying academically gifted students: An aptitude perspective. *Gifted Child Quarterly, 49*(2):111–138.

Lohman, D.F., J.W. Pellegrino, D.L. Alderton, and J.W. Regian. (1987). Dimensions and components of individual differences in spatial abilities. In S.H. Irvine and S.E. Newstead, Eds., *Intelligence and Cognition: Contemporary Frames of Reference* (pp. 253–312). Dordrecht, Holland: Kluwer.

Lubinski, D. (2010). Spatial ability and STEM: A sleeping giant for talent identification and development. *Personality and Individual Differences, 49*(4):344–351.

Malinowski, J.C., and W.T. Gillespie. (2001). Individual differences in performance on a large-scale, real-world wayfinding task. *Journal of Environmental Psychology, 21*(1):73–82.

McHenry, J.J., L.M. Hough, J.L. Toquam, M.A. Hanson, and S. Ashworth. (1990). Project A validity results: The relationship between predictor and criterion domains. *Personnel Psychology, 43*(2):335–354.

Morris, R.G., and D. Parslow. (2004). Neurocognitive components of spatial memory. In G.L. Allen and D. Haun, Eds., *Human Spatial Memory: Remembering Where* (pp. 217–247). Mahwah, NJ: Lawrence Erlbaum Associates.

National Research Council. (2006). *Learning to Think Spatially: GIS as a Support System in the K-12 Curriculum*. Committee on Support for Thinking Spatially: The Incorporation of Geographic Information Science Across the K-12 Curriculum. Geographical Sciences Committee, Board on Earth Sciences and Resources, Division on Earth and Life Sciences. Washington, DC: The National Academies Press.

National Research Council. (2013). *Making the Soldier Decisive on Future Battlefields*. Committee on Making the Soldier Decisive on Future Battlefields. Board on Army Science and Technology, Division on Engineering and Physical Sciences. Washington, DC: The National Academies Press.

Peterson, N.G., T.L. Russell, G. Hallam, L. Hough, C. Owens-Kurtz, K. Gialluca, and K. Kerwin. (1990). Analysis of the experimental predictor battery. In J.P. Campbell and L.M. Zook, Eds., *Building and Retaining the Career Force: New Procedures for Accessing and Assigning Army Enlisted Personnel: Annual Report 1990 FY* (Research Note 952, pp. 73–199). Alexandria, VA: U.S. Army Research Institute for the Behavioral and Social Sciences.

Powers, R. (2013). *ASVAB for Dummies: Premier PLUS*. Hoboken, NJ: Wiley & Sons.

Project Talent Office. (1961). *Project Talent*. Washington, DC: Project Talent Office. Available: http://www.projecttalent.org/docs/Project_Talent_Tech_Manual.pdf [October 2014].

Raven, J.C. (1992). *Raven's Progressive Matrices.* San Antonio, TX: The Psychological Corporation.

Roberts, D.R., G.N. Goff, F. Anjoul, P.C. Kyllonen, G. Pallier, and L. Stankov. (2000). The Armed Services Vocational Aptitude Battery (ASVAB): Little more than acculturated learning (Gc)!? *Learning and Individual Differences, 12*(1):81–103.

Robertson, K.F., S. Smeets, D. Lubinski, and C.P. Benbow. (2010). Beyond the threshold hypothesis: Even among the gifted and top math/science graduate students, cognitive abilities, vocational interests, and lifestyle preferences matter for career choice, performance, and persistence. *Current Directions in Psychological Science, 19*(6):346–351.

Russell, T.L., and N.G. Peterson. (2001). The experimental battery: Basic attribute scores for predicting performance in a population of jobs. In J.P. Campbell and D.J. Knapp, Eds., *Exploring the Limits in Personnel Selection and Classification* (pp. 269–306). Hillsdale, NJ: Lawrence Erlbaum Associates.

Sackett, P.R., M.J. Eitelberg, and W.S. Sellman. (2009). *Profiles of American Youth: Generational Changes in Cognitive Skill.* Alexandria, VA: Human Resource Research Organization.

Shea, D.L., D. Lubinski, and C.P. Benbow. (2001). Importance of assessing spatial ability in intellectually talented young adolescents: A 20-year longitudinal study. *Journal of Educational Psychology, 93*(3):604–614.

Shepard, R.N. (1978). The mental image. *American Psychologist, 33*(2):125–137.

Sims, V.K., and R.E. Mayer. (2002). Domain specificity of spatial expertise: The case of video game players. *Applied Cognitive Psychology, 16*(1):97–115.

Smith, I.M. (1964). *Spatial Ability.* San Diego, CA: Knapp.

Stieff, M., B.L. Dixon, M. Ryu, B.C. Kumi, and M. Hegarty. (2014). Strategy training eliminates sex differences in spatial problem solving in a STEM domain. *Journal of Educational Psychology, 106*(2):390–402.

Terlecki, M.S., N.S. Newcombe, and M. Little. (2008) Durable and generalized effects of spatial experience on mental rotation: Gender differences in growth patterns. *Applied Cognitive Psychology, 22*(7):996–1,013.

Uttal, D.H., N.G. Meadow, E. Tipton, L.L. Hand, A.R. Alden, C. Warren, and N.S. Newcombe. (2013). The malleability of spatial skills: A meta-analysis of training studies. *Psychological Bulletin, 139*(2):352–402.

Voyer, D., S. Voyer, and M.P. Bryden. (1995). Magnitude of sex differences in spatial abilities: A meta-analysis and consideration of critical variables. *Psychological Bulletin, 117*(2):250–270.

Wai, J., D. Lubinski, and C.P. Benbow. (2009). Spatial ability for STEM domains: Aligning over fifty years of cumulative psychological knowledge solidifies its importance. *Journal of Educational Psychology, 101*(4):817–835.

Wang, L., A.S. Cohen, and M. Carr. (2014). Spatial ability at two scales of representation: A meta-analysis. *Learning and Individual Differences, 36*:140–144.

Wolfe, J.H. (1997). Incremental validity of ECAT battery factors. *Military Psychology, 9*(1):49–76.

Section 3

Identification and Prediction of New Outcomes

5

Teamwork Behavior

Committee Conclusion: Research has identified a number of individual-differences attributes that are broadly predictive of success in a team environment. There has also been progress in identifying attributes that when aggregated across team members (e.g., mean level of cognitive ability, minimum agreeableness), are predictive of team effectiveness. More research is needed to expand and amplify this work in the context of potential utility in military accession. The committee concludes that the teamwork knowledge, skills, abilities, and other characteristics (KSAO) domain merits inclusion in a program of basic research with the long-term goal of improving the Army's enlisted accession system.

The small unit has always been critical to an army's success. The U.S. Army's selection of soldiers for assignment into a particular team, squad, and platoon is the basis for much of the soldier's military experience and achievement. There are thousands of military units that serve a wide variety of functions such as combat, medical, aviation, rescue, and support (Dyer et al., 1980). Furthermore, Essens and colleagues (2005) argue that more specialized units will be needed to meet new demands as the Army is tasked to add political and social objectives to more traditional military missions. Today's soldiers are challenged to work in multinational coalitions, joint forces operations, and ad hoc teams with nonroutine tasks. This chapter examines current theory and research on teams, which the committee applies to the Army's organizational level of the small unit, and proposes future research directions that are likely to enhance the unit's collective capacity to perform.

This chapter uses Kozlowski and Ilgen's (2006, p. 79) definition of a team:

> A team can be defined as (a) two or more individuals who (b) socially interact (face-to-face or, increasingly, virtually); (c) possess one or more common goals; (d) are brought together to perform organizationally relevant tasks; (e) exhibit interdependencies with respect to workflow, goals, and outcomes; (f) have different roles and responsibilities; and (g) are together embedded in an encompassing organizational system, with boundaries and linkages to the broader system context and task environment.

About 60 years of research on teams have yielded a significant literature on team processes and team effectiveness (Kozlowski and Ilgen, 2006). This research builds on the small-group literature founded in social psychology and extends McGrath's (1964) Input-Process-Output (I-P-O) heuristic to examine what factors shape team processes, how they interact in efforts to reach team goals, and the types of outcomes these interactions or team processes produce. The I-P-O model generally describes inputs (I) as factors at the individual level (e.g., team member personality), team level (e.g., task structure), and organizational/environmental level (e.g., organizational design). These factors are antecedents that enable, inhibit, or enhance team member interactions. Team processes (P) are generally acknowledged as critical mediators between inputs and team outcomes. They involve interpersonal processes as teams cycle through transition phases (e.g., plans for action) and action phases (Marks et al., 2001). Thus, efforts to improve the selection of potential team members and the composition of teams should focus on how individual differences, considered independently or in combination, relate to team processes and their outcomes. Finally, outcomes (O) are results of team processes that include performance outcomes (e.g., team effectiveness, team efficiency) as well as behavioral (e.g., absenteeism, turnover) and affective outcomes (e.g., team member commitment, team viability) (Cohen and Bailey, 1997; Mathieu et al., 2008). Variations and extensions of the I-P-O model abound, with different emphases on temporal dynamics (Marks et al., 2001), multilevel aspects of I-P-O (Kozlowski and Klein, 2000), and emergent states that serve as additional mediators between inputs and outcomes (Ilgen et al., 2005).

This chapter identifies future research needs that can improve the Army's collective capacity to perform. Specifically, it focuses on how selection and classification of entry-level enlisted soldiers can improve unit performance and mission success. The I-P-O model will serve as a loose framework to identify future research objectives. Starting with the end goal, the committee first discusses team outcomes to define the criteria domain for selection and classification. Next, we examine team processes and emergent states as more proximal criteria of collective capacity. Finally, we examine how future research on individual-level inputs to teams might help understand who is best suited for teamwork and how individuals might be better classified into specific Army small units, including teams, squads, and platoons.

TEAM OUTCOMES: DEFINING TEAM EFFECTIVENESS

Mathieu and Gilson (2012) noted that there has been relatively little research on team outcomes. Defining and measuring team outcomes have been challenging because they are often tied to specific team tasks and organizational conditions. These idiosyncratic measures can limit the generalizability of the research. A review of work team research (Sundstrom et al., 2000) found a wide variety of outcome constructs (e.g., productivity, communication, satisfaction, accidents, prosocial behavior) as well as measures to represent team outcomes, (e.g., objective measures of quantity and quality of team output; aggregated measures of individual ratings on team satisfaction and motivation; and managerial or customer ratings of team overall performance). Production teams were more likely to have objective measures of team performance, whereas service teams were more likely to have subjective self-ratings of team outcomes.

Despite the wide variety of team outcomes, Mathieu and Gilson (2012) identified two general forms. Tangible outcomes are directly related to team goals and include criteria tapping productivity, efficiency and quality, or composites of these outcomes. In contrast, influences on team members are outcomes that include team-level emergent states (e.g., unit cohesiveness) as well as individual-level outcomes tapping attitudes, behaviors, reactions, and individual development related to teamwork. More research attention has been paid to tangible outcomes at the individual role, team, and organization levels (Mathieu et al., 2008); however, team member reactions are also important because they are likely to drive future team interactions and team viability (Hackman, 1990; Mathieu et al., 2008).

Other challenges to defining and measuring these outcomes focus on temporal dynamics and the multilevel nature of team outcomes. Mathieu and colleagues (2008) noted that teams vary on how long it may take to develop stable outcomes. Teams do different things at different times (Marks et al., 2001) and evolve over the course of the team's developmental stages (Tuckman, 1965). LePine (2003) examined how team-level averages of member cognitive ability, achievement, dependability, and openness to change predicted team performance before and after an unforeseen change in the task. None of these team composites predicted routine performance before the change, but significant relationships were found between these predictors, mediated by role structure adaptability, and team performance in a changing context. Thus, research gaps exist in understanding not only *what* defines team effectiveness but also in understanding *when* these outcomes should be measured and over what period of time.

Teams vary on how individual team members' actions are combined into a team-level outcome (Kozlowski et al., in press). Combinations range from *composition* models (e.g., individual team member errors are aggregated to represent team errors) to *compilation* models (e.g., differences in knowledge expertise across team members yield new insights on task goals). In composition models, constructs at different levels are isomorphic, amenable to simple aggregations based on sums or means, and empirically supported by indices of within-group agreement (e.g., r_{wg}; as defined by Bliese, 2000). The performance of an army fire team (e.g., small team of riflemen firing at multiple targets) may be a composition construct if construed as the sum of targets that are hit by any team member. Individual hits and team hits share the same form and function.

In contrast, compilation models involve constructs in a common domain, but differ in their emergence. For example, the performance of an artillery team may be a compilation construct if construed as the number of targets that are hit by the team. In this team context, individual performance differs across team members and must be highly coordinated for team performance to emerge. The fire direction officer checks the target location, a technical expert ensures all equipment is ready, two fire direction specialists coordinate horizontal and vertical operations, and other team members may drive vehicles, operate the radio, chart data, etc.[1] Thus, the form of a team outcome can influence how it is measured and how individual-level actions and characteristics may be combined to understand the team outcome.

Attempts to use individual-level characteristics to directly predict team-level outcomes are likely to result in cross-level and misspecification fallacies (Ployhart and Schneider, 2002). Given the multilevel nature of individuals and teams, selecting individuals to maximize team outcomes can be achieved in two basic ways (Ployhart and Schneider, 2002, 2005). In the first selection model, individual-level KSAOs are used to predict individual-level outcomes that are related to team effectiveness. This is the traditional selection model in human resources management, incorporating team-relevant criteria such as individual-level reactions, attitudes, behavior, and personal development (Mathieu and Gilson, 2012). This approach also requires theoretical and empirical links between individual-level criteria and team-level outcomes. These links are obvious in composition models

---

[1] This example derives from a paper prepared by Captain Andrew Miller, former U.S. Army, for the National Research Council's Committee on the Context of Military Environments: Social and Organizational Factors. The paper is available by request from the public access file of that committee.

but pose research and measurement challenges for compilation models (Kozlowski and Klein, 2000). An example of this approach is a study by Morgeson and colleagues (2005) that found conscientiousness, extraversion, and agreeableness to be significantly related to supervisor ratings of an individual's contextual (team) performance.

The second selection model aggregates individual-level KSAOs to form team-level measures to directly predict team-level outcomes. Chan (1998) described five types of models to combine lower-level data to represent higher-level phenomena: additive (e.g., mean), direct consensus (e.g., within-group consensus of individual perceptions), referent-shift (e.g., within-group consensus of individual perceptions of the team), dispersion (e.g., variance), and process (e.g., focus on process or change). These types designate different functional relationships of the bottom-up process of outcome emergence. With regard to personnel selection, all types may be used to represent team-level outcomes, but only additive and dispersion models are used to represent team-level predictors.

In addition to combinations of individual-level data, team-level constructs may be represented by a single score from one team member. Minimum or maximum scores within a team can describe situations where one individual has a great effect on the entire group (e.g., one bad apple spoils the whole barrel or one brilliant mind carries the whole team) (Day et al., 2004). An example of this approach is a study by Mohammed and Angell (2003) that found student team variability on extraversion was positively correlated with team presentation grades (team oral performance).

These two selection models have direct implications for selection system design. The traditional individual-level approach can be used to measure KSAOs at pre-accession to predict individual-level outcomes related to team performance. These KSAOs can also be used in a multilevel approach to aid classification of individuals, post-accession, into teams whose team profiles are most likely to enhance team effectiveness.

In addition to their influence on team outcomes as described above, team processes may also serve as more proximal criteria for selection purposes. Instead of focusing on how individuals contribute to team performance, one can examine relationships between individual KSAOs on specific teamwork behaviors such as coordination, communication, and conflict resolution. In this approach, team processes and emergent states of teams are presented as critical mediators between individual characteristics and team outcomes.

TEAMWORK PROCESSES AND EMERGENT STATES OF TEAMS

Team outcomes represent bottom-line or distal criteria for selection systems. However, the I-P-O model identifies several processes and emergent states that mediate relationships between individual KSAOs and ultimate tangible outcomes of teams. Consequently, they can serve as selection criteria because many team processes and emergent states have been shown to predict team effectiveness (Mathieu et al., 2008). Marks and colleagues (2001, p. 237) define team processes as "members' interdependent acts that convert inputs to outcomes through cognitive, verbal, and behavioral activities directed toward organizing taskwork to achieve collective goals." Cycles of I-P-O episodes involve transition and action phases, with previous episodes influencing subsequent episodes. Transition phases include team processes related to planning and evaluation activities as a team plans activities for goal accomplishment or reviews an action for lessons learned. In contrast, action phases include processes related to individual and coordinated behaviors that are directly tied to goal attainment. Finally, interpersonal processes occur through planning and action phases, focusing on motivation, conflict, and affect management.

As an example of a model of team processes, Marks and colleagues (2001) describe 10 team processes that are nested under the rubrics of transition, action, and interpersonal processes as follows:

Transition processes:
(1) a team's mission analysis;
(2) goal specification (prioritization of goals and identified subgoals); and
(3) strategy formulation and alternative action plans.

Action processes:
(4) monitoring goal progress;
(5) monitoring environment and resources;
(6) monitoring team members and providing back-up support if needed; and
(7) coordination of individual tasks in an efficient sequence.

Interpersonal processes:
(8) managing conflict within the team;
(9) managing motivation for taskwork; and
(10) managing emotions of team members.

A meta-analysis examining relationships between team processes and two outcomes, team performance and team member satisfaction, showed all team processes were positively related to both team outcomes (LePine et al., 2008). Furthermore, there was some evidence that these relationships were moderated by team size and task interdependence, so that stronger relationships were found for larger teams and teams with high task interdependence. However, the meta-analysis was limited by its domain of a small number of studies that predominantly used paper-and-pencil surveys for data collection.

Marks and colleagues (2001, p. 237) described emergent states as "cognitive, motivational, and affective states of teams, as opposed to the nature of their member interaction." These states develop from dynamic interactions of I-P-O factors within a team context, emerging after early team experiences and changing over time with subsequent experiences. Emergent states may be viewed as team outcomes or as mediators between team inputs and tangible team outcomes (Mathieu et al., 2008). Some states, such as team cognition and team cohesion, may initially be influenced by surface-level individual characteristics (e.g., race or sex), but deeper forms of individual differences (e.g., personality or knowledge) may be better predictors in more mature teams (Kozlowski and Chao, 2012). Like team processes, the emergent states can serve as proximal criteria for team selection and classification decisions.

New technologies should be explored to better assess teamwork behaviors beyond paper-and-pencil measures. For example, current research with sociometric badges (a wearable electronic device about the size of an ID card that measures patterns of behavior) allows researchers to collect real-time data in social networks (Hollingshead and Poole, 2012; Kozlowski, in press; Pentland, 2010). These technologies can capture team interactions with several biomarkers (e.g., physical activity, identity, vocal intensity, heart rate, physical proximity with other team members), tracking who interacts with whom, when, and for how long (Kozlowski, in press). Badges can record not only what was said but also physiological data that may be able to capture qualitative metrics of the conversations (e.g., changes in heart rate during interaction with a particular team member may be interpreted as stress or anxiety). Technological improvements on the capabilities of these badges to record team interactions should increase the reliability and validity of these team-behavior measures.

In another example, computational models can be used to examine the correlation of a wide variety of initial team characteristics with emerging outcomes (McGrath et al., 2000). They simulate teams by specifying mathematical equations (e.g., logical if-then statements) to describe team interactions from one point in time to the next (Kozlowski et al., 2013). Studies using computational modeling can avoid typical constraints of experimental methods such as limited sample sizes, fatigue effects, and restriction of range in subject characteristics. Computational models have been used to identify possible effects of various individual member learning rates on team learning (Kozlowski et al., 2014). These results were combined with experiments on human teams to validate metrics for the emergence of team knowledge. Computational models have also been used to compare different social network measures on leadership influence (Braun et al., 2014).

This method enabled Braun and colleagues to examine over 500,000 simulated teams in a wide range of conditions. Their results showed that social network metrics that were most commonly used in experimental designs (e.g., reciprocity and centralization metrics) were not the best predictors of leadership influence. Although there was no single best network metric for all leadership outcomes, indirect network metrics (e.g., betweenness and closeness metrics) were better predictors of leadership influence (Braun et al., 2014). Given the practical constraints of team research, computational modeling may prove useful in the study of several individual-level KSAOs and their simultaneous effects on team behaviors and emergent team states.

As noted above, there has been more research on team processes than on team outcomes (Mathieu and Gilson, 2012). Thus, there is a need to develop a better understanding of, and new metrics to operationalize, team outcomes and effectiveness. In addition, new technologies should be explored to better assess teamwork behaviors beyond paper-and-pencil measures (see objective A in the research recommendation at the end of this chapter). To identify additional areas for future research, the committee next reviews what is known about individual-level predictors of team processes and outcomes.

TEAM MEMBER INPUTS: SELECTING AND CLASSIFYING INDIVIDUALS FOR EFFECTIVE TEAMS

Early research on military teams identified two primary skill tracks necessary for effective team performance (Glickman et al., 1987; Morgan et al., 1986; see also Shuffler et al., 2012 for a review of teams in military environments[2]). *Taskwork* requires specific job-related knowledge, skills, and abilities that are directly tied to performance demands of the job. They are bound by job requirements at the individual level. Traditional selection systems focus on assessing how well candidates can perform individual taskwork. In contrast, *teamwork* addresses the coordinated efforts of team members as they work together to accomplish individual and team goals. Teams that clearly understand each member's role, communicate well, and have members who support each other are likely to be more effective than teams without good teamwork. Thus, proficiency in both taskwork and teamwork, operating together, is necessary for teams to be efficient and effective. Although the importance of teamwork has been widely recognized

---

[2] Note that Glickman and colleagues (1987) and Morgan and colleagues (1986) both considered Navy teams, whereas Shuffler and colleagues (2012) provide a review of research relevant to teams across military environments and services.

in the team's literature, research on selection for teamwork is not well developed (Mohammed et al., 2010; Tannenbaum et al., 2012).

Employee selection at the individual level generally includes examination of job analyses, identification of KSAO predictors, defining criteria, and measurement issues regarding the reliability and validity of specific selection procedures. Selection for team members may be viewed with parallel features at multiple levels. Although teams are defined by the interactions of team members, some taskwork may be performed at the individual level (Arthur et al., 2005). Thus, taskwork can be a blend of individual and coordinated work efforts. Job analysis generally defines taskwork at the individual level, whereas team task analysis examines the criticality of team task interdependence from both taskwork and teamwork perspectives (Bowers et al., 1994; Mohammed et al., 2010). Recent work has identified *team-relatedness*, the extent to which team members must interact in order to maximize team effectiveness, and team workflow among team members as important components of team task analysis (Arthur et al., 2012). However, this area is relatively undeveloped, and more research is needed to define team tasks (Allen and West, 2005).

Reviews on team selection have generally identified demographics (e.g., race), task-related KSAOs (e.g., experience), and psychological individual differences (e.g., cognitive ability, personality) as predictors of team selection and classification (Allen and West, 2005; Mohammed et al., 2010; Morgeson et al., 2012). Current research on team composition (the process by which individuals are selected for and assigned to a team, including consideration of potential and existing team members' individual characteristics) has examined the diversity of team members on a wide range of individual characteristics. Reviews examining surface-level or demographic diversity found null (Horwitz and Horwitz, 2007) or negative (Mannix and Neale, 2005) relationships between this type of team composition and team outcomes. Extensions of this line of research have examined multiple characteristics that may define subgroups, such as group fault lines (Lau and Murnighan, 1998), with stronger fault lines related to team conflict (Lau and Murnighan, 2005). However, a meta-analysis found the negative effects of team demographic diversity on team performance diminished over time (Bell et al., 2007).

In contrast, reviews of team composition based on deeper-level individual differences such as general intelligence, personality, or values showed significant effects on team performance (Bell, 2007; Stewart, 2006). Team-level operationalizations of team composition were generally means on a single individual difference, although other measures (e.g., maximum score for disjunctive tasks, minimum score for conjunctive tasks) have been used with less frequency. Most robust was the finding that team means on general mental ability were positively related to team performance (Bell, 2007;

Stewart, 2006). Bell (2007) also found team means on conscientiousness, openness to experience, collectivism, and preference for teamwork were positively related to team performance in field studies but not in laboratory studies. In addition, team minimum values on agreeableness were stronger predictors of team performance than other operationalizations of a team agreeableness composite. Bell called for future research to explore additional operationalizations (e.g., proportion of team with high conscientiousness), as well as possible combinations (e.g., mean and maximum) to better represent team composition.

More recently, Carton and Cummings (2013) examined the number of subgroups based on social identities (i.e., surface-level) and knowledge (i.e., deeper-level). They found that having just two subgroups based on social identities had a more negative impact on team performance than having either no subgroups or more than two subgroups. However, having more knowledge-based subgroups generally had a positive effect on team performance. Furthermore, teams performed better when identity-based subgroups were not balanced in size but performed better when knowledge-based subgroups were balanced. Thus, conclusions about the effects of team composition on team performance must take into account the individual-level variables that are used to compose the team as well as the number and relative sizes of subgroups. The meta-analyses also show that relationships between team composition and team performance are moderated by factors such as research setting (field or laboratory studies), operationalization of team-level variables (Bell, 2007), and task type (Stewart, 2006). Furthermore, most of the current research examines only one or a few individual differences. Future research is needed to identify individual and team cognitions, affect/motivation, and behaviors that are linked to successful team outcomes and effectiveness. A team task analysis that, for example, identifies critical generic KSAOs, such as core teamwork skills, could be useful for initial selection (pre-accession). It could also aid in classifying individuals, post-accession, with respect to the more-contingent teamwork competencies identified as important for the proficiency of high-value teams. Essential to this research area is developing methods of team task analysis (see objective B of the research recommendation at the end of this chapter).

In addition to effective team composition, successful teams can be described as having individuals who are experienced and skilled in teamwork. Mohammed and colleagues (2010) described teamwork skills, such as interpersonal skills and communication skills, as core teamwork competencies that can be measured at the individual level and used to aid team selection. These skills would be valuable to all types of teams, making them generic predictors of teamwork. Often assessed by a work sample or paper-and-pencil test, specific teamwork skills like adaptability skills (Salas et al., 2007), interpersonal skills (Morgeson et al., 2005), and communication skills (Bowers et al., 2000; Smith-Jentsch et al., 1996) have been shown to be valid predictors of team outcomes.

Stevens and Campion (1994) developed a Teamwork Knowledge, Skills, and Ability (KSA) test to help select individuals who are suited to teamwork; however, results on the validation of this test as a selection tool are mixed (Allen and West, 2005). Self-reported KSAs of teamwork were significantly related to individual performance (McClough and Rogelberg, 2003), but mean Teamwork KSA scores were not related to team performance (Miller, 2001). Furthermore, Teamwork KSA scores were found to be significantly correlated with cognitive ability (Stevens and Campion, 1999), potentially limiting the utility of this predictor if added to an existing test battery that already includes general mental abilities (Miller, 2001).

This report describes a number of individual differences that may be predictors of an individual's ability to work in teams. For example, an individual's inhibitory control capacity (see Chapter 2), cognitive biases (see Chapter 3), and emotional regulation (see Chapter 6) may be related to how well he or she adapts when team members engage in potentially stressful interactions. A soldier who is capable of controlling emotional and behavioral impulses may be more likely to work well in a team context. Conversely, a soldier who is low in emotional regulation may be likely to disrupt or distract the team from accomplishing a mission.

Individual assessments may be combined to help predict a number of outcomes related to long-term team performance and satisfaction. For selection and placement decisions, it is likely that there are no simple rules to find the best individuals for a particular team. Individual characteristics related to teamwork and taskwork may provide supplementary fit to a team (e.g., all team members are similarly conscientious and responsive to one another); or the characteristics may provide complementary fit (e.g., one team member's expertise fills a team's need for that knowledge).

In addition to core teamwork competencies that apply to all teams, contingent teamwork competencies are sensitive to team tasks, structures, and environmental conditions that may change relationships between predictors and outcomes (Mohammed et al., 2010). A meta-analysis of studies examining *person-group fit* found this individual-level construct to be significantly correlated with individual-level outcomes such as job satisfaction, organizational commitment, and intentions to quit (Kristof-Brown et al., 2005). Unfortunately, the small number of studies in this meta-analysis (n's ranged from 4 to 12, depending on outcome) did not permit any investigation of possible moderators to these relationships. Perceptions of fit based on goals, values, and/or personality are likely to be influenced by the existing team and organization environments that a newly selected team member joins. For example, team size will influence the division of labor and coordination demands (Steiner, 1972). An individual's perceptions of fit within a team can be shaped by these contextual features. As stated in objective C of the research recommendation at the end of this chapter, future research is needed to identify optimal within-individual profiles that are linked to team effectiveness. This research should also consider types of team structures, tasks, and environmental conditions that moderate relationships between profile attributes and their combined influence on team processes and outcomes.

In addition to potential moderators that can change predictor-outcome relationships, it is possible for post-selection experiences to change the predictive power of individual profiles. Pre-selection experiences help shape an applicant's task-related and team-related knowledge and skills, and they can be used as predictors of future team outcomes. Similarly, post-selection experiences help shape an employee's task-related and team-related knowledge and skills, potentially mitigating the utility of a selection measure. Training programs and team experiences can increase an individual's capacity toward effective teamwork. For example, Chen and colleagues (2004) found a course on teamwork significantly improved Teamwork KSA scores and observer ratings of teamwork competencies for college students. Thus, training may compensate for low pre-selection scores on this predictor. In another example, team leadership may require adjustments when a team encounters an extreme context that puts members in harm's way (Rumsey, 2013; Yammarino et al., 2010). As a team assesses a situation, plans for action, and executes those plans, the team learns how it impacts the environment. In turn, these lessons influence subsequent teamwork (Burke et al., 2006). Future research should investigate the effects of teamwork training and team experiences on the predictive power of individual-differences measures (see objective D in the research recommendation at the end of this chapter).

A PATH FORWARD

Research Gaps and Future Directions

Teams are critical units for military performance. Improving team performance can be aided by selecting individuals who are most capable of teamwork and composing teams with individuals who have compatible KSAOs. The criterion domain of team effectiveness has received relatively little research attention compared to research on team inputs and team processes. A better understanding of team outcomes, both tangible performance metrics and influences on team members, is needed to validate team selection methods. Well-defined team outcomes can also inform more-thorough descriptions and analyses of team tasks. Despite repeated calls for more research on team task analysis, few researchers have answered this call (Allen and West, 2005).

Team selection can use traditional selection models, finding individual-level predictors (e.g., cognitive ability) to predict individual-level outcomes (e.g., individual performance). Team selection can also take a multilevel perspective, examining links between individual- and team-level predictors and their relationships to individual- and team-level criteria. Some important team characteristics (e.g., team cohesiveness, team diversity) have no individual-level equivalents, so multilevel perspectives are better able to assess a team's collective capacity to perform. Indeed, Ployhart and Schneider (2005) argued that some desired organizational characteristics (e.g., workforce diversity) may not be achieved if selection systems only focus on maximizing individual performance on a single job. Research on team-level composites of predictors in team selection can help identify new predictors at the individual level and how they might be best combined to measure team composition.

It is important to recognize the limitations of team selection. The benefits of a valid selection system may be nullified if team members fail to cooperate with one another (Schneider et al., 2000). Good selection can identify those individuals who are most likely to succeed in teams; however, the actual interactions of team members in a specific context would be more directly responsible for team outcomes (Hackman and Katz, 2010). What happens post-accession—how individuals are trained, equipped, organized, socialized, led, and rewarded—will also be important predictors of how team members interact and perform, but this topic was beyond the scope of this study. Likewise, the composition or assembly of individuals into teams brings together a wide variety of individual characteristics, team task designs, and contextual features that can critically affect team performance. Furthermore, different team members can assume "leadership" roles as individuals mature, members are reassigned, and time changes role demands (Contractor et al., 2012). Teamwork behaviors may also be influenced by such negative factors as stereotypes or implicit bias, but the committee judges that the utility of using such predictors for selection purposes would be limited due to potential mitigating effects of contextual features such as good leadership and clear tasks or of surface-level diversity whereby the effects of race or gender differences dissipate over time as team members get to know one another (Harrison et al., 2002). Lastly, the committee notes that a collective capacity to perform can be defined by larger units than teams as defined in this chapter. The Army's squads, platoons, companies, battalions, brigades, divisions, corps, and even field army extend the "collective capacity" to many levels. Since this chapter is focused on individual selection, it makes sense to confine future research at individual and team levels. However, as Ployhart and Schneider (2002) noted, some higher-level goals may require lower-level goals to be suboptimized in order to accommodate performance requirements across all levels.

Relative to individual personnel selection, research on team selection is in a developmental stage. More research is needed to identify KSAOs that are required for specific team taskwork as well as generic teamwork. More research is needed to define and measure team effectiveness as teams develop, evolve, and change. More research is needed to identify individual-level predictors, how they are combined into individual profiles, and how they are combined into team composites. Together, these research directions can maximize the potential for individuals to work effectively in dynamic teams.

RESEARCH RECOMMENDATION

The U.S. Army Research Institute for the Behavioral and Social Sciences should support research on individual- and team-level knowledge, skills, abilities, and other characteristics that influence the collective capacity to perform. Future research should include the following objectives:

A. Develop a better understanding of, and new metrics to operationalize, team outcomes and effectiveness. In addition, new technologies should be explored to better assess teamwork behaviors beyond paper-and-pencil measures.
B. Identify individual and team cognitions, affect/motivation, and behaviors that are linked to successful team outcomes and effectiveness. Essential to this is developing methods of team task analysis.
C. Identify optimal within-individual profiles that are linked to team effectiveness. This research should also consider types of team structures, tasks, and environmental conditions that moderate relationships between profile attributes and their combined influence on team processes and outcomes.
D. Investigate the effects of teamwork training and team experiences on the predictive power of individual-differences measures.

REFERENCES

Allen, N.J., and M.A. West. (2005). Selection for teams. In A. Evers, N. Anderson, and O. Voskuijl, Eds., *The Blackwell Handbook of Personnel Selection* (pp. 476–494). Malden, MA: Wiley-Blackwell.

Arthur, W., Jr., B.D. Edwards, S.T. Bell, A.J. Villado, and W. Bennett, Jr. (2005). Team task analysis: Identifying tasks and jobs that are team-based. *Human Factors*, 47(3):654–669.

Arthur, W., Jr., R.M. Glaze, A. Bhupatkar, A.J. Villado, W. Bennett, Jr., and L.J. Rowe. (2012). Team task analysis: Differentiating between tasks using team relatedness and team workflow as metrics of team task interdependence. *Human Factors*, 54(2):277–295.

Bell, S.T. (2007). Deep-level composition variables as predictors of team performance: A meta-analysis. *Journal of Applied Psychology*, 92(3):595–615.

Bell, S.T., A.J. Villado, M.A. Lukasik, A. Briggs, and L. Belau. (2007). *Revisiting Team Demographic Diversity and Performance Relationship: A Meta-Analysis*. Paper presented at the Annual Conference of the Society for Industrial and Organizational Psychology, New York, NY.

Bliese, P.D. (2000). Within-group agreement, non-independence, and reliability: Implications for data aggregation and analysis. In K.J. Klein and S.W.J. Kozlowski, Eds., *Multilevel Theory, Research, and Methods in Organizations: Foundations, Extensions, and New Directions* (pp. 349–381). San Francisco, CA: Jossey-Bass.

Bowers, C.A., D.P. Baker, and E. Salas. (1994). Measuring the importance of teamwork: The reliability and validity of job/task analysis indices of team-training design. *Military Psychology*, 6(4):206–214.

Bowers, C.A., J.A. Pharmer, and E. Salas. (2000). When member homogeneity is needed in work teams: A meta-analysis. *Small Group Research*, 31(3):305–327.

Braun, M.T., L.A. DeChurch, and D.R. Carter. (2014). *Measuring Shared Leadership: A Computational Modeling Study*. Paper presented at the 29th Annual Conference of the Society for Industrial and Organizational Psychology, Honolulu, HI.

Burke, C.S., K.C. Stagl, E. Salas, L. Pierce, and D. Kendall. (2006). Understanding team adaptation: A conceptual analysis and model. *Journal of Applied Psychology*, 91(6):1,189–1,207.

Carton, A.M., and J.N. Cummings. (2013). The impact of subgroup type and subgroup configurational properties on work team performance. *Journal of Applied Psychology*, 98(5):732–758.

Chan, D. (1998). Functional relations among constructs in the same content domain at different levels of analysis: A typology of composition models. *Journal of Applied Psychology*, 83(2):234–246.

Chen, G., L.M. Donahue, and R.J. Klimoski. (2004). Training undergraduates to work in organizational teams. *Academy of Management Learning and Education*, 3(1):27–40.

Cohen, S.G., and D.E. Bailey. (1997). What makes teams work: Group effectiveness research from the shop floor to the executive suite. *Journal of Management*, 23(3):239–290.

Contractor, N.S., L.A. DeChurch, J. Carson, D.R. Carter, and B. Keegan. (2012). The topology of collective leadership. *The Leadership Quarterly*, 23(6):994–1,011.

Day, E.A., A. Winfred, B. Miyashiro, B.D. Edwards, T.C. Tubre, and A.H. Tubre. (2004). Criterion-related validity of statistical operationalizations of group general cognitive ability as a function of task type: Comparing the mean, maximum, and minimum. *Journal of Applied Psychology*, 34(7):1,521–1,549.

Dyer, J.L., T.R. Tremble, Jr., and D.L. Finley. (1980). *The Structural, Training, and Operational Characteristics of Army Teams* (Technical Report 507). Alexandria, VA: U.S. Army Research Institute for the Behavioral and Social Sciences.

Essens, P., A. Vogelaar, J. Mylie, C. Blendell, C. Paris, S. Halpin, and J. Baranski. (2005). *Military Command Team Effectiveness: Model and Instrument for Assessment and Improvement* (Research and Technology Organisation Technical Report TR-HFM-087). NATO. Available: http://www.dtic.mil/cgi-bin/GetTRDoc?AD=ADA437898 [February 2015].

Glickman, A.S., S. Zimmer, R.C. Montero, P.J. Guerette, W.J. Campbell, B.B. Morgan, and E. Salas. (1987). *The Evolution of Team Skills: An Empirical Assessment with Implications for Training* (Technical Report NTSC 87-016). Orlando, FL: Naval Training Systems Center, Human Factors Division.

Hackman, J.R., Ed. (1990). *Groups that Work (and Those that Don't): Creating Conditions for Effective Teamwork*. San Francisco, CA: Jossey-Bass.

Hackman, J.R., and N. Katz. (2010). Group behavior and performance. In S.T. Fiske, G.T. Gilbert, G. Lindzey, and A.E. Jongsma, Eds., *Handbook of Social Psychology* (Vol. 5, pp. 1,208–1,251). New York: Wiley & Sons.

Harrison, D.A., K.H. Price, J.H. Gavin, and A.T. Florey. (2002). Time, teams, and task performance: Changing effects of surface- and deep-level diversity on group functioning. *Academy of Management Journal*, 45(5):1,029–1,045.

Hollingshead, A.B., and M.S. Poole. (2012). *Research Methods for Studying Groups and Teams: A Guide to Approaches, Tools, and Technologies*. New York: Routledge.

Horwitz, S.K., and I.B. Horwitz. (2007). The effects of team diversity of team outcomes: A meta-analytic review of team demography. *Journal of Management*, 33(6):987–1,015.

Ilgen, D.R., J.R. Hollenbeck, M. Johnson, and D. Jundt. (2005). Teams in organizations: From input-process-output models to IMOI models. *Annual Review of Psychology*, 56:517–543.

Kozlowski, S.W.J. (in press). Advancing research on team process dynamics: Theoretical, methodological, and measurement consideration. *Organizational Psychology Review*. Available: http://iopsych.msu.edu/koz/Recent%20Pubs/Kozlowski%20%28in%20 press%29%20-%20Team%20Dynamics.pdf [February 2015].

Kozlowski, S.W.J., and G.T. Chao. (2012). The dynamics of emergence: Cognition and cohesion in work teams. *Managerial and Decision Economics*, 33(5-6):335–354.

Kozlowski, S.W.J., and D.R. Ilgen. (2006). Enhancing the effectiveness of work groups and teams. *Psychological Science*, 7(3):77–124.

Kozlowski, S.W.J., and K.J. Klein. (2000). A multilevel approach to theory and research in organizations: Contextual, temporal, and emergent processes. In K.J. Klein and S.W.J. Kozlowski, Eds., *Multilevel Theory, Research, and Methods in Organizations* (pp. 3–90). San Francisco, CA: Jossey-Bass.

Kozlowski, S.W.J., G.T. Chao, J.A. Grand, M.T. Braun, and G. Kuljanin. (2013). Examining the dynamics of multilevel emergence: Learning and knowledge building in teams. *Organizational Research Methods*, 16(4):581–615.

Kozlowski, S.W.J., G.T. Chao, J.A. Grand, M.T. Braun, G. Kuljanin, D. Pickhardt, and S. Mak. (2014). *Team Macrocognition: Integrating Computational Modeling and Research Methods*. Paper presented at the 29th Annual Conference of the Society for Industrial and Organizational Psychology, Honolulu, HI.

Kozlowski, S.W.J., J.A. Grand, S.K. Baard, and M. Pearce. (in press). Teams teamwork, and team effectiveness: Implications for human systems integration. In D. Boehm-Davis, F. Durso, and J. Lee, Eds., *The Handbook of Human Systems Integration*. Washington, DC: American Psychological Association.

Kristof-Brown, A.L., R.D. Zimmerman, and E.C. Johnson. (2005). Consequences of individuals' fit at work: A meta-analysis of person-job, person-organization, person-group, and person-supervisor fit. *Personnel Psychology*, 58(2):281–342.

Lau, D.C., and J.K. Murnighan. (1998). Demographic diversity and faultlines: The compositional dynamics of organizational groups. *Academy of Management Review*, 23(2):25–34.

Lau, D.C., and J.K. Murnighan. (2005). Interactions within groups and subgroups: The effects of demographic faultlines. *Academy of Management Journal*, 48(4):645–659.

LePine, J.A. (2003). Team adaptation and postchange performance: Effects of team composition in terms of members' cognitive ability and personality. *Journal of Applied Psychology*, 88(1):2–39.

LePine, J.A., R.F. Piccolo, C.L. Jackson, J.E. Mathieu, and J.R. Saul. (2008). A meta-analysis of teamwork processes: Tests of a multidimensional model and relationships with team effectiveness criteria. *Personnel Psychology, 61*(2):273–307.

Mannix, E., and M.A. Neale. (2005). What differences make a difference? *Psychological Science in the Public Interest, 6*(2):31–55.

Marks, M.A., J.E. Mathieu, and S.J. Zaccaro. (2001). A temporally based framework and taxonomy of team processes. *Academy of Management Review, 26*(3):356–376.

Mathieu, J.E., and L.L. Gilson. (2012). Criteria issues and team effectiveness. In S.W.J. Kozlowski, Ed., *The Oxford Handbook of Organizational Psychology* (pp. 910–932). New York: Oxford University Press.

Mathieu, J., M.T. Maynard, T. Rapp, and L. Gilson. (2008). Team effectiveness 1997–2007: A review of recent advancements and a glimpse into the future. *Journal of Management, 34*(3):410–476.

McClough, A.C., and S.G. Rogelberg. (2003). Selection in teams: An exploration of the teamwork knowledge, skills, and ability test. *International Journal of Selection and Assessment, 11*(1):56–66.

McGrath, J.E. (1964). *Social Psychology: A Brief Introduction*. New York: Holt, Rinehart, and Winston.

McGrath, J.E., H. Arrow, and J.L. Berdahl. (2000). The study of groups: Past, present, and future. *Personality and Social Psychology Review, 4*(1):95–105.

Miller, D.L. (2001). Reexamining teamwork KSAs and team performance. *Small Group Research, 32*(6):745–766.

Mohammed, S., and L.C. Angell. (2003). Personality heterogeneity in teams: Which differences make a difference for team performance? *Small Group Research, 34*(6):651–677.

Mohammed, S., J. Cannon-Bowers, and F.S. Chuen. (2010). Selection for team membership: A contingency and multilevel perspective. In J.L. Farr and N.T. Tippins, Eds., *Handbook of Employee Selection* (pp. 801–822). New York: Routledge.

Morgan, B.B., A.S. Glickman, E.A. Woodard, A.S. Blaiwes, E. Salas, W.J. Campbell, D.L. Miller, R.C. Montero, and S. Zimmer. (1986). *Measurement of Team Behaviors in a Navy Training Environment* (NTSC Technical Report No. TR-86-014). Orlando, FL. Human Factors Division, Naval Training Systems Command, Department of the Navy. Available: http://www.dtic.mil/dtic/tr/fulltext/u2/a185237.pdf [January 2015].

Morgeson, F.P., M.H. Reider, and M.A. Campion. (2005). Selecting individuals in team settings: The importance of social skills, personality characteristics, and teamwork knowledge. *Personnel Psychology, 58*(3):83–611.

Morgeson, F.P., S.E. Humphrey, and M.C. Reeder. (2012). Team selection. In N. Schmitt, Ed., *The Oxford Handbook of Personnel Assessment and Selection* (pp. 832–848). New York: Oxford University Press.

Pentland, A.S. (2010). To signal is human: Real-time data mining unmasks the power of imitation, kith and charisma in our face-to-face social networks. *American Scientist, 98*:203–211.

Ployhart, R.E., and B. Schneider. (2002). A multi-level perspective on personnel selection research and practice: Implications for selection system design, assessment, and construct validation. In F.J. Yammarino and F. Dansereau, Eds., *Research in Multi-level Issues*, Vol. 1, (pp. 95–140). Oxford, UK: Elsevier Science.

Ployhart, R.E., and B. Schneider. (2005). Multilevel selection and prediction: Theories, methods, and models. In A. Evers, N. Anderson, and O. Voskuijl, Eds., *The Blackwell Handbook of Personnel Selection* (pp. 495–516). Malden, MA: Blackwell.

Rumsey, M. (2013). Military understanding of teams. In G.B. Graen and J.A. Graen, Eds., *Management of Team Leadership in Extreme Context: Defending Our Homeland, Protecting Our First Responders*. Charlotte, NC: Information Age.

Salas, E., D.R. Nichols, and J.E. Driskell. (2007). Testing three team training strategies in intact teams: A meta-analysis. *Small Group Research, 38*(4):471–488.

Schneider, B., D.B. Smith, and W.P. Sipe. (2000). Personnel selection psychology: Multi-level considerations. In K.J. Klein and S.W.J. Kozlowski, Eds., *Multi-level Theory, Research, and Methods in Organizations: Foundations, Extensions, and New Directions* (pp. 91–120). San Francisco, CA: Jossey-Bass.

Shuffler, M.L., D. Pavlas, and E. Salas. (2012). Teams in the military. In J.H. Laurence and M.D. Matthews, Eds., *The Oxford Handbook of Military Psychology* (pp. 282–310). New York: Oxford University Press.

Smith-Jentsch, K.A., E. Salas, and D.P. Baker. (1996). Training team performance-related assertiveness. *Personnel Psychology, 49*(4):909–936.

Steiner, I.D. (1972). *Group Processes and Productivity*. New York: Academic Press.

Stevens, M.J., and M.A. Campion. (1994). The knowledge, skill, and ability requirements for teamwork: Implications for human resource management. *Journal of Management, 20*(2):503–530.

Stevens, M.J., and M.A. Campion. (1999). Staffing work teams: Development and validation of a selection test for teamwork settings. *Journal of Management, 25*(2):207–228.

Stewart, G.L. (2006). A meta-analytic review of relationships between team design features and team performance. *Journal of Management, 32*(1):29–54.

Sundstrom, E., M. McIntyre, T. Halfhill, and H. Richards. (2000). Work groups: From the Hawthorne studies to work teams of the 1990s and beyond. *Group Dynamics: Theory, Research, and Practice, 4*(1):44–67.

Tannenbaum, S.I., J.E. Mathieu, E. Salas, and D. Cohen. (2012). Teams are changing: Are research and practice evolving fast enough? *Industrial and Organizational Psychology: Perspectives on Science and Practice, 5*(1):2–24.

Tuckman, B.W. (1965). Developmental sequence in small groups. *Psychological Bulletin, 63*:384–399.

Yammarino, F.J., M.D. Mumford, M.S. Connelly, and S.D. Dionne. (2010). Leadership and team dynamics for dangerous military contexts. *Military Psychology, 22*(Suppl 1):S15–S41.

Section 4

Hybrid Topics with Joint Focus on New Predictor Constructs and Prediction of New Outcomes

6

Hot Cognition: Defensive Reactivity, Emotional Regulation, and Performance under Stress

Committee Conclusion: "*Hot cognition*" *includes the topics of defensive reactivity, emotional regulation, and performance under stress. Research and military experience suggest that the ability to perform well in situations that elicit emotional responses is important in many contexts that are relevant to the military. Research on performance has tended to underplay the role emotions can play in governing behavior, whether for good or bad. The committee concludes that the hot cognition domain merits inclusion in a program of basic research with the long-term goal of improving the Army's enlisted accession system.*

This chapter considers three topics—defensive reactivity, emotional regulation, and performance under stress—that share a common theme of being concerned with one's ability to function well in situations that elicit strong emotions. Researchers have called this ability "hot cognition" (Abelson, 1963; Brand, 1987) to contrast it with the arguably better understood and more commonly researched topic of cognition under circumstances of cool, level, or moderate emotions, or cold cognition.

Topics covered in this chapter overlap somewhat with topics covered in some of the other chapters. Hot cognition is often related to biased judgments and decision making, such as motivated reasoning, and therefore overlaps with topics covered in Chapter 3 on cognitive and decision biases. For example, Loewenstein (2007) distinguished defense mechanisms (e.g., denial, projection, rationalization), which he suggested were automatic and unconscious, from affect regulation behaviors (reappraisal, distraction/

suppression of thoughts and feeling), which he proposed were deliberate processes.

Hot cognition also tends to be concerned with executive functioning and therefore overlaps with Chapter 2 on working memory. Specifically, Zelazo and colleagues (2010) contrasted hot versus cool executive function, with cool executive function referring to "conscious goal-directed problem solving" and hot executive function referring to motivated cognition. An older research tradition from an individual-differences perspective contributes to our current understanding of the moderating effects of emotions, particularly the effects of anxiety on performance (e.g., Humphreys and Revelle, 1984; Hembree, 1988; Byron and Khazanchi, 2011). Current understanding of hot cognition additionally draws on work in social psychology because hot cognition is often a social phenomenon. It also draws on developmental psychology because of the role emotions play in the thinking and judgment related to development. Neuroscience is featured in many studies of hot cognition.

In a presentation given at the public workshop hosted by this committee (National Research Council, 2013), Christopher Patrick proposed a psychometrically oriented approach to the study of individual differences in hot-cognitive processes. This approach, which he called psychoneurometrics, seeks to develop reliable neurobiological assessments of trait constructs such as inhibitory control (see Chapter 2, final section) or defensive reactivity by combining differing known biomarkers of the target trait as assessed psychometrically into a composite neurometric index of the trait (see Figure 2-5 in Chapter 2). Examples of known biomarkers of inhibitory control include the P3 and error-related negativity brain responses; a well-established biomarker of dispositional defensive reactivity is fear-enhanced startle (see Chapter 10 and Appendix C for further discussion of biomarkers). Patrick suggested that this psychoneurometric approach, which uses trait-related biomarkers as "items" to form a neurobiological "scale" can (a) lead to new conceptions of traits and assessments that predict physiological (including brain) reactivity in performance contexts more effectively than report-based measures; (b) minimize rating-scale response bias through the use of physiological measures; and (c) encourage a process-level understanding of constructs, focusing on the biological mechanisms mediating the stimulus-response link.

This chapter is divided into the three topic areas; for each, the committee defines the topic, discusses some of the key findings in the area, reviews how topic-related constructs are typically measured, and then concludes with a discussion of whether a continued research program is justified, based on findings and research prospects. The chapter ends with the committee's recommended future basic research agenda on topics of hot cognition.

DEFENSIVE REACTIVITY/FEARFULNESS VERSUS FEARLESSNESS/BOLDNESS

Defensive reactivity, which can be defined as "proneness to negative emotional reactivity in the face of threat" (as presented by Christopher Patrick, see National Research Council, 2013, p. 23), is related to the construct of fearfulness versus fearlessness or boldness.[1] It is distinguishable from the Big Five personality domain of Neuroticism (or negative emotional stability) at least conceptually in that it entails variations in cue-specific fear reactivity, rather than free-floating negative affect, as Neuroticism is presumed to entail. There is some evidence (e.g., Dvorak-Bertscha et al., 2009; Gordon et al., 2004; Kramer et al., 2012) that while situation cues are relevant in understanding defensive reactivity, important individual differences exist that can be measured, thereby suggesting potential utility for military selection. Specifically, these differences are related to the sensitivity or responsiveness of the brain's defense system (e.g., the amygdala and related structures). Defensive reactivity/fearfulness versus fearlessness/boldness is considered to be a general factor that operates across social, sensation-seeking, and reported-affect domains.

Key Findings

Why is defensive reactivity important? Fear can be an unproductive emotion across many situations ranging from the battlefield to the classroom. Its dispositional opposite, fearlessness or boldness, can be a productive emotional attribute. Boldness has been found to relate to adaptability. For example, Dvorak-Bertscha and colleagues (2009) found that individuals scoring high on boldness were able to maintain their attention on a task under conditions of shock threat (adaptability). There is also some evidence that boldness is a trait useful in leadership. Lilienfeld and colleagues (2012) used Big Five trait ratings of U.S. presidents provided by expert historical biographers to estimate scores on "fearless dominance" or boldness as a facet of psychopathy, and they found this dimension to be related to leadership ability. Among the presidents assessed, they found that Theodore Roosevelt scored the highest on this factor. Elsewhere, Lykken (1995) characterized Winston Churchill as another political leader exhibiting extreme dispositional fearlessness.

There is as yet little research on the relationship between dispositional defensive reactivity and performance in enlisted military occupational spe-

---

[1] While it could be that individuals might be rated separately on fearfulness (high versus low) and boldness (high versus low), current research treats the terms interchangeably. Furthermore, the committee found it hard to imagine a highly fearful individual who is nevertheless highly bold, or a fearless person who is not also bold.

cialties such as infantry, ordnance disposal, security police, or special forces. There might be times when, and situations in which, boldness could be a productive and adaptive attribute. There might be other times when, taken too far, boldness might be maladaptive—for example, with combat soldiers unnecessarily risking their lives and the lives of others by disregarding safety protocol or danger signals. An analysis of optimal ranges of boldness for different military occupational specialties could be a useful and productive endeavor. The context of environments and situations experienced by soldiers across Army specialties might be a challenge to generating optimal levels of boldness for use in the selection process; however, the committee believes an exploration of the application of the Yerkes-Dodson law[2] would also likely be an informative and useful endeavor.

Measures

An important issue is how to measure fearfulness versus boldness. The most common measurement approach has been the use of surveys (rating-scale self-reports). A comprehensive empirical study of fearlessness surveys suggested three distinct categories of scales: social behavior scales, activity preference scales, and perceived experience scales (Kramer et al., 2012). Examples of social behavior scales (which from a content perspective bear a resemblance to the Big Five domain of Extraversion) are the Social Potency (later renamed Social Influence) scale from the Psychopathic Personality Inventory (PPI) and the Shyness with Strangers vs. Gregariousness scale of the Tridimensional Personality Questionnaire (TPQ).[3] These scales are clearly aligned with the Extraversion facets of Social Dominance (e.g., "have leadership abilities") and Gregariousness (e.g., "love to chat"). (Note: In these and other examples of scale items in this chapter, the actual items from the scale are proprietary, so the committee has provided illustrative items from the International Personality Item Pool [Goldberg et al., 2006], based on item or scale content matches.)

Examples of activity preference scales within the Sensation-Seeking domain are the Fearlessness scale of the PPI (e.g., "enjoy the thrill of fearful situations") and the Thrill and Adventure Seeking scale of the Sensation Seeking Scale (e.g., "seek adventure"). Another related scale is the Fear of Uncertainty vs. Confidence scale of the TPQ ("face danger confidently").

---

[2] The Yerkes-Dodson law indicates an increasing linear relationship between arousal and performance, up until a certain point beyond which further increases in arousal have a detrimental effect on performance.

[3] The latter scale is abbreviated as "TPQ-HA3" in the literature because it is a lower-order scale for the higher order Harm Avoidance factor, which Cloninger (1987, p. 575) described as a "heritable tendency to respond intensely to signals of aversive stimuli, thereby learning to inhibit behavior to avoid punishment, novelty, and frustrative non-reward."

Examples in the perceived experience category are the Fearlessness scale of the EAS [Emotionality, Activity, and Sociability] Temperament Survey ("would fear walking in a high-crime part of a city," negatively keyed), the Anticipatory Worry vs. Uninhibited Optimism scale of the TPQ ("often worry about things that turn out to be unimportant"), the Fatigability and the Asthenia vs. Vigor scales of the TPQ ("get too tired to do anything"), and the Stress Immunity scale of the PPI ("recover quickly from stress and illness"). Yet another example of a perceived experience scale is the Fear Survey schedule of the PPI, in which participants rate a series of words such as "flying insects" and "sight of knives" on a *0* to *4* scale to indicate the degree to which the objects described invoke fear (Tomlin et al., 1984). Furthermore, as described in this report's first chapter, some of the constructs related to defensive reactivity are measured through the Tailored Adaptive Personality Assessment System (TAPAS; Drasgow et al., 2012).

An obvious problem with survey-based measures is that in the form of rating scales they are easily faked. So, in addition to the survey approach there has been some research that explored actual performance-behavior measures. An intriguing approach is to measure fearfulness physiologically, for instance as increased startle-blink reactivity to sudden loud noises occurring in fear-evoking situations (the "fear-potentiated startle" that indicates a fear reaction). A study by Vaidyanathan and colleagues (2009) showed a significant correlation between degree of startle potentiation during aversive picture viewing and survey responses related to fear versus boldness (see also Kramer et al., 2012). Such measures might not be practical for large-scale personnel testing using present day technology, but the point of this illustration is to show that performance measures are possible, and it may be productive to devise and develop performance-based measures of the kinds of constructs that have been identified from survey-based research.

Potential Benefits of Future Research

The committee sees potential for improvements in identifying candidates likely to succeed as soldiers through the pre-accession assessment of defensive reactivity (fearfulness versus boldness). Research suggests that boldness can contribute to adaptability and leadership (see Chapter 7 for a more detailed discussion of adaptability as a predictor construct), whereas fear can be an unproductive emotion across many situations ranging from the battlefield to the classroom. The committee also believes there is high potential for improved tests resulting from an exploration of both physiological measures and performance measures of defensive reactivity such as, but not limited to, the eye-blink startle measure. More generally, it may be useful to develop performance-based measures of personality factors relevant to fear/fearlessness, such as the Big Five domains of Extraversion and Neuroticism, which are traditionally measured with surveys. The committee believes that a research program along these lines would contribute to fundamental knowledge of how biobehavioral dispositions such as defensive reactivity relate to and differ from the Big Five domains, would clarify the predictive validity (for various outcomes) of survey versus performance measures, and would identify potential contextual factors (e.g., boldness in social versus affective versus workplace versus battlefield contexts) and evaluate their importance.

EMOTION REGULATION

Consider these situations in which a person regulates (or fails to regulate) his or her emotions: suppressing the impulse to seek revenge for unfair treatment, resisting a temptation, experiencing anxiety, or "acting out." Emotion regulation refers to the "cognitive and behavioral processes that influence the occurrence, intensity, duration, and expression of emotion" (Campbell-Sils and Barlow, 2007, p. 543). Individuals can regulate their emotions by intensifying them or by denying, weakening, curtailing, masking, or completely hiding them. Emotion regulation can be seen as a form of coping with situations by modifying one's emotional reactions; a way of increasing or decreasing the intensity of the moment (Gross, 2002).

Gross and Thompson (2007) proposed a *modal model* of emotion, involving a situation that is attended to and appraised, resulting in a response. In their scheme, emotions are one kind of affect along with stress, mood, and impulses. Emotion regulation applies to both positive and negative emotions, can be conscious (controlled) or unconscious (automatic) or somewhere in between, and can be good or bad, depending on the situation. The constructs of coping with stress, mood regulation, and psychological defenses overlap with emotion regulation but are not the same. Coping with stress is addressed in the next section of this chapter; psychological defenses were discussed in Chapter 3 on cognitive biases. Mood regulation is addressed here, to a limited extent.

Another process or stage model of emotions was proposed by Siegler and colleagues (2006), who distinguished internal feeling states (i.e., the subjective experience of emotion), emotion-related cognitions (e.g., thought reactions to a situation), emotion-related physiological processes (e.g., heart rate, hormonal, or other physiological reactions), and emotion-related behavior (e.g., actions or facial expressions related to emotion).

The related construct of self-regulation has been defined as the "process by which one monitors, directs attention, maintains, and modifies behaviors to approach a desirable goal" (Ilkowska and Engle, 2010, p. 266). As Gross and Thompson (2007) pointed out, self regulation and emotion regulation might share processes, as can be seen in delay-of-gratification studies. For example, Mischel's (1996) marshmallow studies asked respondents (typically children) to ignore a marshmallow on display in front of them while the experimenter steps out of the room, with the promise of a reward of two marshmallows in a short time when the experimenter returns if the respondent has resisted the temptation and failed to consume the marshmallow. Such studies call on attentional responses—reframing and distraction—which are involved in emotion regulation as well as in pain regulation and other forms of self-regulation. Open issues are whether and to what extent findings and methods for researching self-regulation overlap with those for emotion regulation. (Self-regulation is covered in Chapter 2 on working memory.)

Individuals differ in the way in which they can or typically do regulate their emotions, which suggests that there are emotion-regulation *abilities* (Lopes et al., 2005). This idea is incorporated into the Mayer-Salovey-Caruso definition of emotional intelligence (Mayer et al., 2002) as a facet called emotional management (along with three other facets: emotional understanding, perceiving emotions, and facilitating emotions). More importantly, MacCann and colleagues (2014) provided evidence that emotional intelligence behaves from a psychometric standpoint as a second-stratum factor of abilities, alongside fluid, crystallized, and other abilities in the Carroll (1993) abilities taxonomy. A further subset of emotional management, called "controlled interpersonal affect regulation," refers to the deliberate regulation of someone else's affect (Niven et al., 2009). Emotional management may be related to the Big Five domain of Neuroticism (Diefendorff and Richard, 2003).

The point here is not that a particular existing test of emotional intelligence, such as the Mayer-Salovey-Caruso Emotional Intelligence Test (MSCEIT) should be included in the Armed Services Vocational Aptitude Battery (ASVAB) but rather that the constructs such tests endeavor to measure may be useful. Certainly there are many challenges associated with measuring such constructs in a fair, valid, and reliable way in order to evaluate whether they increase the prediction of Army outcomes beyond what is already given by existing cognitive and personality assessments.

Why is emotion regulation important? People regulate their emotions for a variety of purposes, including to avoid pain or get pleasure (hedonic motivation), to conform to social roles, to facilitate task or role performance, to manage self-presentation, and to regulate the feelings of others. Emotional coping with adversity ("flying off the handle") is counterproductive. Emotional regulation may be related to self-discipline, which is key to success in school and the workplace (see, for example, Duckworth, 2011).

Key Findings

Emotional management (as measured by the Situational Test of Emotion Management, described below) has been found to correlate (r = .54) with eudaimonic well-being (the sense of living a "meaningful life") as measured by a rating scale and with hedonic well-being (experiences of happiness or pleasure) as measured by a diary method (the Day Reconstruction Method). Correlations between net affect—that is positive affect minus negative affect while engaged in an activity—ranged from r = .18 for sleeping and resting to r = .44 for working (Burrus et al., 2012). Although well-being outcomes are not often included in discussions of what makes soldiers successful, they may mediate outcomes such as attrition and therefore are worth noting.

There is some evidence that coping styles, which are related to emotion regulation, are related to achievement. MacCann and colleagues (2012) found that, after controlling for personality, cognitive ability, and demographic factors, problem-based coping predicted grades, life satisfaction, and positive feelings about school. Emotion-focused coping was found to predict negative feelings only. Avoidant coping predicted both positive and negative feelings about school. Emotional intelligence and problem-focused coping have been proposed as inoculators against fatigue from the effort in showing compassion to others (Zeidner, 2013).

Matthews and colleagues (2006) compared performance in stressful versus non-stressful tasks and found that low emotional intelligence was correlated with worry states and avoidance coping, after controlling for personality factors. However, emotional intelligence was not specifically related to task-induced changes in stress state. Neuroticism correlated with distress, worry, and emotion-focused coping. Conscientiousness correlated with task-focused coping.

Another finding on emotion regulation pertains to development. The improvement of emotion regulation through the course of adolescence has been attributed to maturation of the frontal lobes, which are essential for controlling attention and inhibiting thoughts and behaviors (Siegler, 2006). Developmental trends in emotion regulation are particularly important given that the average Army enlistment age is 20 years of age.[4] Recruits' likely growth in emotion-regulation skills as they develop from adolescents into adults may have implications for selection and for prediction of both near-term and long-term performance.

---

[4] See the Army's frequently asked questions about recruiting for more information. Available: http://www.usarec.army.mil/support/faqs.htm#age [July 2014].

Measures

There are several measures of emotional regulation used in the literature. The MSCEIT includes a Managing Emotion subscale (Mayer et al., 2003), which is defined as "the ability to be open to feelings, to modulate them in oneself and others so as to promote personal understanding and growth." The two tests in the Managing Emotion category are a social management test and a personal management test. In both tests, participants read a story and rate (on a scale from *very effective* to *very ineffective*) how effective various responses would be to handle the emotions in the story.

The Situational Test of Emotional Management (MacCann and Roberts, 2008) consists of 44 situational judgment tests (see Chapter 9 for further discussion of situational judgment tests) with items such as

> Lee's workmate fails to deliver something on time, causing Lee to fall behind schedule. What's most effective? (a) work harder to compensate; (b) get angry with the workmate; (c) explain the urgency of situation; or (d) never rely on that workmate again.

The Coping with School Situations Questionnaire (MacCann et al., 2011) measures various coping styles (problem-focused, emotion-focused, and avoidant coping strategies) with survey-type questions. For example, agreement with "I make the extra effort to get all my activities completed" indicates a problem-focused strategy; agreement with "I blame myself for having put off my homework" indicates an emotion-focused strategy; and agreement with "When faced with a test the next day I go out with my friends" indicates an avoidant strategy (MacCann et al., 2012).

Other measures commonly seen in the literature (which could test emotion regulation with the addition of emotional stimuli in the task) include the flanker test (Eriksen and Eriksen, 1974), a measure of attentional control in which respondents are shown a stimulus to which they are to respond (e.g., by indicating whether it is pointing to the left or to the right, or whether it is a noun or a verb, or whether it is a square or a circle), but the respondent is simultaneously shown distracting visual stimuli, which they are told to ignore. Related tests are the Stroop color-word test and the Simon spatial compatibility tasks. The outcome measure is a difference in performance with and without the distractor present (or with congruent versus incongruent distractors). With adults, the distracting information is designed to influence attention, and the measure therefore becomes one of attentional control (e.g., Shaffer and LaBerge, 1979). Flanker tests have been used successfully with children as young as 4 or 5 years (Diamond et al., 2007). Emotion-related flanker tests using faces have also been developed for adults (Fenske and Eastwood, 2003).

Another commonly used technique is the marshmallow test: the delay-of-gratification test described above. With respect to executive function, certain problem-solving tasks are considered cool executive functioning tasks, such as the Wisconsin Card Sort (which asks respondents to match a new card to categories, but the categories sometimes change), whereas others, such as the Iowa Gambling task (Damásio, 2008), which has participants choose cards to win money from card decks that vary in their payout probabilities, are thought to recruit hot executive functioning. Both types of task see phenomena such as perseverance behavior (staying on the classification strategy or on the same card deck) and are executive function tasks, but the former is thought to involve the prefrontal cortex, while the latter engages the dorsolateral prefrontal cortex.

Potential Benefits of Future Research

The committee sees potential for improvements in military selection and assignment tests through a pre-accession assessment of emotion regulation. Emotion regulation is important in a broad variety of contexts, ranging from controlling one's impulses (e.g., the impulse to seek revenge on a party blamed for some perceived slight or injustice) to recovering from a loss or stressful event or coping in a clear-headed manner with a catastrophic situation. Clearly, emotion regulation is important in military operations, particularly combat or any deployment. Emotion regulation comes into play in controlling impulses to obey command authority and follow the law of land warfare. It is called upon in recovering from combat-related loss, such as in post-traumatic stress and post-traumatic stress disorder, and is likely related to vulnerability and conveying resilience. Third, emotion regulation is likely to be an important factor in combat effectiveness and leadership in combat and more generally during war experiences.

Emotion regulation has emerged in the literature as a construct of importance, as shown, for example, in the increasing number of references to it (14 total before 1990; 603 in the 1990s; 2,785 in 2001–2005) and by the publishing of a handbook focused on it (Gross, 2007). There is widespread agreement on some of the key features of emotion regulation. For example, emotion-regulation strategies include situation selection and modification (e.g., avoiding situations that might provoke negative emotions); attentional deployment, such as distraction or redirecting focus; reappraisal (reinterpreting an event, "when life gives you lemons, make lemonade"); and response modulation, such as suppressing emotions (e.g., to appear appropriate) or intensifying them (e.g., to gain sympathy; to invoke fear in others) through such means as manipulating one's facial expressions or body posture (Bargh and Williams, 2007; Gross, 2007; Mesquita and Albert, 2007).

There also have been significant relevant developments in affective neuroscience, including understanding of the physiological basis for emotional regulation, such as the neural mechanisms of inhibition (Quirk, 2007), the role of the prefrontal cortex in higher cognitive control including affective processing (Davidson et al., 2007), and the role of the dorsal region of the anterior cingulate cortex in monitoring conflicts—not only in tasks such as the Stroop, Simon, and flanker tasks but also between emotional and cognitive influences on certain tasks such as moral dilemmas (McClure et al., 2007). Much is still unknown, but it seems rapid advances are occurring. Continued investment in this topic area will likely pay dividends in increased clarity of how emotions and cognition together affect human capability and performance.

There also may be benefit to pursuing more traditional measures: rating scales, situational judgments, and performance measures of emotion regulation. Situational judgment tests (see Chapter 9) seem quite amenable to presenting and recording responses to emotion-provoking situations (e.g., situations presented via video) that require an emotion-regulation strategy and response, particularly if the response as acted out by the respondent is recorded in detail (e.g., captured on video). Among promising performance measures, an example would be an adult version of Mischel's (1996) "marshmallow test" as a delay-of-gratification measure. Such tasks are referred to as intertemporal choice tasks, and they pit impulsivity against patience by asking the subject to choose between an immediate small award or a delayed but greater reward (McClure et al., 2007). Other performance measures of interest would include other kinds of conflict tests, such as variants of the Simon, flanker, and Stroop tasks, which set up a conflict between two response tendencies. In an emotion-regulation variant of these tests, the response conflict might be between a more emotional and a more level-headed cognitive response. Another category of performance measures could include moral dilemma tests and variants on tasks from game theory, such as the prisoner's dilemma and the ultimatum game. Such approaches enable the manipulation of respondents' perceptions of unfairness and injustice, provoking an emotional reaction and allowing for an emotional response (e.g., retribution).

Finally, for emotion regulation in contrast to other constructs in the individual-differences literature, while it is well know that emotion regulation develops over a lifetime, there is less known about the degree to which emotion-regulation skills are trainable and transferable from situation to situation. The literature on emotional intelligence (Wranik et al., 2007) posits emotion management or emotion regulation as one "branch" of emotional intelligence and treats it as a trait-like factor. But very little is yet known about the extent to which emotion regulation is malleable. Given the importance of emotion regulation to so much of human activity, the committee believes that exploring not only the testing of emotional regulation but also its trainability is an important topic for future research on human capability.

PERFORMANCE UNDER STRESS

Performance pressure has been defined as an anxious desire to perform at a high level in a given situation (Hardy et al., 1996). The difference between regular performance and pressure performance depends at least somewhat on the subjective importance of a situation (Baumeister, 1984; Beilock and Carr, 2001). The related construct of *choking*, or performing more poorly than expected, given one's level of skill, tends to occur in situations fraught with performance pressure, especially in sensorimotor or action-based skills (e.g., basketball free throws; golf putting) (Beilock and Carr, 2001).

Performance under stress is clearly important in military selection because there are many situations in the workplace and battlefield that involve working under pressure, including time pressure, pressure in the context of someone evaluating a person's performance, or pressure in the context of danger or risk up to and including the high-level risk of death in combat operations. The consequences of mistakes in these pressure-to-perform situations can be catastrophic to the individual and to those who constitute the individual's small unit.

The ability to perform well under stress could be considered a performance outcome, but it may also be a general attribute of a person (e.g., in sports there is the concept of a "clutch" performer) or it could be to some extent situational. It does not have to be one or the other. It is for this reason that this chapter is included in the hybrid topics section of the report. In psychometrics, particularly clearly in Item Response Theory models (e.g., the Rasch model), there are separable concepts of ability and item difficulty. The stressfulness of a situation can be seen as an item difficulty construct whereas an individual's general ability to handle stressful situations can be seen as an ability construct. There is the additional issue, from a psychometrics perspective of dimensionality, that situations can potentially vary not only in their stressfulness overall but in the kinds of stressfulness they provide. Anxiety inducements—such as time pressure or reputation—and financial incentives, drugs, fatigue, unfamiliar surroundings, noise, and other factors (e.g., Evans and Cohen, 1987) can in some sense be seen as stressors. It is likely that different stressors affect individuals differentially.

Key Findings

The ability to perform under stress can be treated as an outcome that conventional personality measures can to some extent already predict. For example, the U.S. Army Research Institute for the Behavioral and Social Sciences (ARI) has Assessment of Background and Life Experiences (White et al., 2001), Assessment of Individual Motivation (Stark et al., 2011) and TAPAS (Drasgow et al., 2012) measures, which include scales for adjustment and other indicators of emotional stability. There is research suggesting that such measures do predict performance-under-stress outcomes (e.g., Drasgow et al., 2012). In addition, there is ARI research on the psychological (i.e., an Occupational Stress Assessment Inventory) and psychophysiological (e.g., heart rate, vagal tone, blood pressure) predictors of performance under stress (Heslegrave and Colvin, 1996).

Scientific understanding of performance under stress has advanced over the past decade and a half. One lesson learned is that conscious attention to proceduralized skills promotes choking, suggesting that it is often better "not to think about it," a skill that itself can be trained (Beilock and Carr, 2001). One way "not to think about it" is to engage simultaneously in a task unrelated to the task at hand. For example, Beilock and colleagues (2002) showed that expert soccer players could dribble a ball as well when repeating random words as they could without distractions. However, their dribbling performance was impaired when they had to perform a task related to the dribbling, such as signaling whether they had just used the inside or outside of their foot to move the ball. The related task seems to have encouraged (or forced) them to "think about it." Understanding the conditions promoting or preventing choking and other maladaptive performances under stress is important in developing an assessment framework for measuring susceptibility to performance stressors. Not taking into account issues such as these could result in the development of tests measuring different constructs for different people.

Pressure to perform, such as that induced by a monetary incentive, generally diminishes performance on a cognitive (e.g., math) test (Beilock and Carr, 2005). The proposed mechanism is that pressure reduces working memory capacity (see Chapter 2). If working memory capacity is reduced, the individual relies on strategies that are less working-memory-intensive, such as perceptual strategies or "intuitive" strategies. By contrast, when tasks are set up to favor an intuitive strategy, pressure improves performance. This difference was demonstrated with a classification task that could be solved with rule learning or with perception, supporting "distraction theory" (Markman et al., 2006).

Measures

A battery of performance-under-stress measures does not yet exist, but there are some standard techniques designed to induce pressure, many of which are inherent in the procedures for administering current standard tests of cognitive and other abilities. One standard technique is to ask participants to perform cognitive tasks (e.g., math, problem solving, classification) under no pressure and under pressure (the form of pressure may be financial incentives, time limits, or the presence of an audience or partner), much like the current procedures for administering the ASVAB or any other standardized test of cognitive ability. Another is to have participants perform athletic tasks (e.g., golf putting, basketball free throws; soccer dribbling) under pressure (e.g., financial incentives, time pressure, audience pressure, or the pressure caused by self-monitoring of performance), much like the current procedures used to assess the physical capability of potential military recruits.

Potential Benefits of Future Research

The committee sees great potential for the identification of future successful soldiers through the assessment of performance under stress. In the past decade there have been significant advances in understanding how cognition changes (e.g., choking) in various pressure-inducing contexts (Beilock et al., 2002; Markman et al., 2006); However, there would be value in further clarifying or characterizing those tasks for which performance changes are most likely and those contexts that induce pressure. For example, what are the characteristics of tasks that make them susceptible to the choking effect? Is the effect limited to perceptual-motor tasks or even more specifically to certain types of tasks (e.g., golf putting; shooting free throws), or is the effect broader? Does it extend to more purely cognitive tasks (test taking), or are the mechanisms underlying choking on cognitive tasks (test anxiety) fundamentally different from those underlying choking on athletic tasks ("the yips," a loss of fine motor skills)? And what is the nature of the pressure that produces this effect? Are time pressure, financial incentives, and peer pressure interchangeable, or do they induce different kinds of effects? What other kinds of pressure-inducing contexts are there?

Finally, what are the salient characteristics of the choking phenomenon, broadly speaking, as an individual-differences construct? Do individuals systematically vary in their susceptibility to pressure? How broad does the construct need to be to capture the behavioral phenomena? Do those whose performance suffers under pressure on athletic tasks also experience test anxiety? What kind of mediating role does experience or expertise play in susceptibility to choking? Are there individual-differences factors that moderate that relationship, such as emotion-regulation ability, Neuroticism, or boldness?

RESEARCH RECOMMENDATION

The U.S. Army Research Institute for the Behavioral and Social Sciences should support research to understand issues in the domain of hot cognition:

A. Research should explore behavioral performance measures and also physiological measures of dispositional defensive reactivity, such as the eye-blink startle measure and other biological indicators (biomarkers) of fear activation, and more generally other traits conceived as "biobehavioral." Research should examine how biobehavioral dispositions like defensive reactivity relate to and are distinct from other personality constructs such as the Big Five (Openness, Conscientiousness, Extraversion, Agreeableness, and Neuroticism). In addition, research should compare the predictive validity of trait dispositions as assessed by physiological or behavioral measures in relation to survey assessments and examine how traits affect performance outcomes in differing situational contexts (e.g., impact of dispositional boldness on behavioral effectiveness in social versus affective versus workplace versus battlefield context).

B. Research should clarify how emotions and cognitions together affect human capability and performance and should expand understanding of the physiological bases for emotional regulation. Key themes include neural mechanisms of inhibition, the role of the prefrontal cortex in higher cognitive control including affective processing, and the role of the dorsal region of the anterior cingulate cortex in monitoring conflicts (e.g., conflict between emotional and cognitive influences on moral dilemma tasks).

C. Research should explore measuring emotional regulation with established forms of assessment such as rating scales, situational judgment tests, and performance measures (e.g., delay-of-gratification measures, emotional conflict tests, cooperation versus competition tasks).

D. Research should examine the conditions that improve or diminish cognition and performance under stress, in order to develop measures of susceptibility to stress.

E. Research should evaluate whether susceptibility to stress is contingent on the type of stressor (e.g., time pressure, peer pressure, fatigue) and whether there are cognitive, personality, and experiential correlates of susceptibility.

REFERENCES

Abelson, R.P. (1963). Computer simulation of "hot cognition." In S.S. Tomkins and S. Messick, Eds., *Computer Simulation of Personality: Frontier of Psychological Theory* (pp. 277–302). New York: Wiley & Sons.
Bargh, J.A., and L.E. Williams. (2007). On the nonconscious regulation of emotion. In J. Gross, Ed., *Handbook of Emotion Regulation* (pp. 429–445). New York: Guilford Press.
Baumeister, R.F. (1984). Choking under pressure: Self-consciousness and paradoxical effects of incentives on skillful performance. *Journal of Personality and Social Psychology,* 46(3):610–620.
Beilock, S.L., and T.H. Carr. (2001). On the fragility of skilled performance: What governs choking under pressure? *Journal of Experimental Psychology: General,* 130(4):701–725.
Beilock, S.L., and T.H. Carr. (2005). When high-powered people fail: Working memory and "choking under pressure" in math. *Psychological Science,* 16(2):101–105.
Beilock, S.L., T.H. Carr, C. MacMahon, and J.L. Starkes. (2002). When paying attention becomes counterproductive: Impact of divided versus skill-focused attention on novice and experienced performance of sensorimotor skills. *Journal of Experimental Psychology: Applied,* 8(1):6–16.
Brand, A.G. (1987). Hot cognition: Emotion and writing behavior. *Journal of Advanced Composition,* 6:5–15.
Byron, K., and S. Khazanchi. (2011). A meta-analytic investigation of the relationship between state and trait anxiety to performance on figural and verbal creative tasks. *Personality and Social Psychology Bulletin,* 37(2):269–283.
Burrus, J., A. Betancourt, S. Holtzman, J. Minsky, C. MacCann, and R.D. Roberts. (2012). Emotional intelligence relates to well-being: Evidence from the situational judgment test of emotional management. *Applied Psychology: Health and Well-Being,* 4(2):151–166.
Carroll, J.B. (1993). *Human Cognitive Abilities: A Survey of Factor Analytic Studies.* New York: Cambridge University Press.
Campbell-Sills, L., and D.H. Barlow. (2007). Incorporating emotion regulation into conceptualizations of treatments of anxiety and mood disorders. In J.J. Gross, Ed., *Handbook of Emotion Regulation* (pp. 542–560). New York: Guilford Press.
Cloninger, C.R. (1987). A systematic method for clinical description and classification of personality variants. A proposal. *Archives of General Psychiatry,* 44(6):573–588.
Damásio, A.R. (2008). *Descartes' Error: Emotion, Reason and the Human Brain.* New York: Random House.
Davidson, R.J., A. Fox, and N.H. Kalin. (2007). Neural bases of emotion regulation in nonhuman primates and humans. In J.J. Gross, Ed., *Handbook of Emotion Regulation* (pp. 47–68). New York: Guilford Press.
Diamond, A., W.S. Barnett, J. Thomas, and S. Munro. (2007). Preschool program improves cognitive control. *Science,* 318(5,855):1,387–1,388.
Diefendorff, J.M., and E.M. Richard. (2003). Antecedents and consequences of emotional display rule perceptions. *Journal of Applied Psychology,* 88(2):284–294.
Drasgow, F., S. Stark, O.S. Chernyshenko, C.D. Nye, C.L. Hulin, and L.A. White. (2012). *Development of the Tailored Adaptive Personality Assessment System (TAPAS) to Support Army Selection and Classification Decisions* (Technical Report No. 1311). Fort Belvoir, VA: U.S. Army Research Institute for the Behavioral and Social Sciences.
Duckworth, A.L. (2011). The significance of self-control. *Proceedings of the National Academy of Sciences of the United States of America,* 108(7):2,639–2,640.
Dvorak-Bertscha, J.D., J.J.Curtin, T.J. Rubinstein, and J.P. Newman. (2009). Psychopathic traits moderate the interaction between cognitive and affective processing. *Psychophysiology,* 46(5):913–921.

Eriksen, B.A., and C.W. Eriksen. (1974). Effects of noise letters upon identification of a target letter in a non-search task. *Perception and Psychophysics, 16*(1):143–149.

Evans, G.W., and S. Cohen. (1987). Environmental stress. In D. Stokols and I. Altman, Eds., *Handbook of Environmental Psychology* (Vol. 1, pp. 571–610). New York: Wiley & Sons.

Fenske, M.J., and J.D. Eastwood. (2003). Modulation of focused attention by faces expressing emotion: Evidence from flanker tasks. *Emotion, 3*(4):327–343.

Goldberg, L.R., J.A. Johnson, H.W. Eber, R. Hogan, M.C. Ashton, C.R. Cloninger, and H.G. Gough. (2006). The international personality item pool and the future of public-domain measures. *Journal of Research in Personality, 40*(1):84–96.

Gordon, H.L., A.A. Baird, and A. End. (2004). Functional differences among those high and low on trait measure of psychopathy. *Biological Psychiatry, 56*(7):516–521.

Gross, J.J. (2002). Emotion regulation: Affective, cognitive, and social consequences. *Psychophysiology, 39*(3):281–291.

Gross, J.J., Ed. (2007). *Handbook of Emotion Regulation*. New York: Guilford Press.

Gross, J.J., and R.A. Thompson. (2007). Emotion regulation: Conceptual foundations. In J.J. Gross, Ed., *Handbook of Emotion Regulation* (pp. 3–20). New York: Guilford Press.

Hardy, L., R. Mullen, and G. Jones. (1996). Knowledge and conscious control of motor actions under stress. *British Journal of Psychology, 87*(4):621–636.

Hembree, R. (1988). Correlates, causes, effects, and treatment of test anxiety. *Review of Educational Research. 58*(1):47–77.

Heslegrave, R.J., and C. Colvin. (1996). *An Exploration of Psychological and Psychophysiological Measures as Predictors of Successful Performance Under Stress* (Technical Report Number 1035). Alexandria, VA: U.S. Army Research Institute for the Behavioral and Social Sciences.

Humphreys, M.S., and W. Revelle. (1984). Personality, motivation, and performance: A theory of the relationship between individual differences and information processing. *Psychological Review, 91*(2):153–184.

Ilkowska, M., and R.W. Engle. (2010). Working memory capacity and self-regulation. In R.H. Hoyle, Ed., *Handbook of Personality and Self-Regulation* (pp. 265–290). West Sussex, UK: Wiley.

Kramer, M.D., C.J. Patrick, R.F. Krueger, and M. Gasperi. (2012). Delineating physiologic defensive reactivity in the domain of self-report: Phenotypic and etiologic structure of dispositional fear. *Psychological Medicine, 42*(6):1,305–1,320.

Lilienfeld, S.O., I.D. Waldman, K. Landfield, A.L. Watts, S. Rubenzer, and T.R. Faschingbauer. (2012). Fearless dominance and the U.S. presidency: Implications of psychopathic personality traits for successful and unsuccessful political leadership. *Journal of Personality and Social Psychology, 103*(3):489–505.

Lopes, P., P. Salovey, M. Beers, and S. Cote. (2005). Emotion regulation abilities and the quality of social interaction. *Emotion, 5*(1):113–118.

Loewenstein, G. (2007). Affect regulation and affective forecasting. In J.J. Gross, Ed., *Handbook of Emotion Regulation* (pp. 180–203). New York: Guilford Press.

Lykken, D. (1995). *The Antisocial Personalities*. Hillsdale, NJ: Lawrence Erlbaum Associates.

MacCann, C., and R.D. Roberts (2008). New paradigms for assessing emotional intelligence: Theory and data. *Emotion, 8*(4):540–551.

MacCann, C., G.J. Fogarty, M. Zeidner, and R.D. Roberts. (2011). Coping mediates the relationship between emotional intelligence (EI) and academic achievement. *Contemporary Educational Psychology, 36*(1):60–70.

MacCann, C., A.A. Lipnevich, J. Burrus, and R.D. Roberts. (2012). The best years of our lives? Coping with stress predicts school grades, life satisfaction, and feelings about high school. *Learning and Individual Differences, 22*(2):235–241.

MacCann, C., D.L. Joseph, D.A. Newman, and R.D. Roberts. (2014). Emotional intelligence is a second-stratum factor of human cognitive abilities: Evidence from hierarchical and bifactor models. *Emotion, 14*(2):358–374.

Markman, A.B., W.T. Maddox, and D.A. Worthy. (2006). Choking and excelling under pressure. *Psychological Science, 17*(11):944–948.

Matthews, G., A.K. Emo, G. Funke, M. Zeidner, R. Roberts, P.T. Costa, Jr., and R. Schulze. (2006). Emotional intelligence, personality, and task-induced stress. *Journal of Experimental Psychology: Applied, 12*(2):96–107.

Mayer, J.D., P. Salovey, and D.R. Caruso. (2002). *Mayer-Salovey-Caruso Emotional Intelligence Test (MSCEIT)*. North Tonawanda, NY: Multi-Health Systems.

Mayer, J.D., P. Salovey, D.R. Caruso, and G. Sitarenios. (2003). Measuring emotional intelligence with the MSCEIT V2.0. *Emotion, 3*(1):97–105.

McClure, S.M., M.M. Botvinick, N. Yeung, J.D. Greene, and J.D. Cohen. (2007) Conflict monitoring in cognition-emotion competition. In. J.J. Gross, Ed., *Handbook of Emotion Regulation* (pp. 204–228). New York: Guilford Press.

Mesquita, B., and D. Albert. (2007). The cultural regulation of emotions. In. J.J. Gross, Ed., *Handbook of Emotion Regulation* (pp. 486–503). New York: Guilford Press.

Mischel, W. (1996). From good intentions to willpower. In P. Gollwitzer and J. Bargh, Eds., *The Psychology of Action* (pp. 197–218). New York: Guilford Press.

National Research Council. (2013). *New Directions in Assessing Performance Potential of Individuals and Groups: Workshop Summary*. R. Pool, Rapporteur. Committee on Measuring Human Capabilities: Performance Potential of Individuals and Collectives. Board on Behavioral, Cognitive, and Sensory Sciences. Division of Behavioral and Social Sciences and Education. Washington, DC: The National Academies Press.

Niven, K., P. Totterdell, and D. Holman. (2009). A classification of controlled interpersonal affect regulation strategies. *Emotion, 9*(4):498–509.

Quirk, G.J. (2007). Prefrontal-amygdala interactions in the regulation of fear. In J.J. Gross, Ed., *Handbook of Emotion Regulation* (pp. 27–46). New York: Guilford Press.

Shaffer, W.O., and D. LaBerge. (1979). Automatic semantic processing of unattended words. *Journal of Verbal Learning and Verbal Behavior, 18*(4):413–426.

Siegler, R., J.S. DeLoache, and N. Eisenberg. (2006). *How Children Develop, Exploring Child Develop Student Media Tool Kit and Scientific American Reader to Accompany How Children Develop*. New York: Worth.

Stark, S., O.S. Chernyshenko, F. Drasgow, W.C. Lee, L.A. White, and M.C. Young. (2011). Optimizing prediction of attrition with the U.S. Army's Assessment of Individual Motivation (AIM). *Military Psychology, 23*(2):180–201.

Tomlin, P., B.A. Thyer, G.C. Curtis, R. Nesse, O. Cameron, and P. Wright. (1984). Standardization of the fear survey schedule based upon patients with DSM-III anxiety disorders. *Journal of Behavioral Therapy and Experimental Psychiatry, 15*(2):123–126.

Vaidyanathan, U., C.J. Patrick, and E.M. Bernat. (2009). Startle reflex potentiation during aversive picture viewing as an index of trait fear. *Psychophysiology, 46*(1):75–85.

White, L.A., M.C. Young, and M.G. Rumsey. (2001). ABLE implementation issues and related research. In J.P. Campbell and D.J. Knapp, Eds., *Exploring the Limits of Personnel Selection and Classification* (pp. 525–558). Mahwah, NJ: Lawrence Erlbaum Associates.

Wranik, T., L.F. Barrett, and P. Salovey. (2007). Intelligent emotion regulation: Is knowledge power? In J.J. Gross, Ed., *Handbook of Emotion Regulation* (pp. 393–407). New York: Guilford Press.

Zeidner, M., D. Hadar, G. Matthews, and R.D. Roberts. (2013). Personal factors related to compassion fatigue in health professionals. *Anxiety, Stress, and Coping, 26*(6):595–609.

Zelazo, P.D., L. Qu, and A.C. Kesek. (2010). Hot executive function: Emotion and the development of cognitive control. In S.D. Calkins and M.A. Bell, Eds., *Child Development at the Intersection of Emotion and Cognition* (pp. 97–111). Washington, DC: American Psychological Association.

7

Adaptability and Inventiveness

Committee Conclusion: The military has a strong interest in adaptive behavior, expressed in terms of assessing novel problems and solving them or acting upon them effectively. Research indicates two promising lines of inquiry. The first would use measures of frequency and quality of ideas generated in open-ended tasks, which have demonstrated incremental validity over and above measures of general cognitive ability for predicting important outcomes related to work performance. The second line of inquiry would use narrow personality constructs to predict adaptive behavior and inventive/creative problem solving. Thus, the committee concludes that idea generation measures and narrow personality measures specific to adaptability and inventiveness merit inclusion in a program of basic research with the long-term goal of improving the Army's enlisted accession system.

BACKGROUND, DEFINITIONS, AND ISSUES

It is essential for organizations seeking to thrive and prosper in a variety of environments to have members who respond effectively to challenging and changing situations whose context may be broad (e.g., interactions with other organizations in the turbulence of international politics) or within the confines of the organization itself (e.g., dealing with coworkers on team projects under constant stress and turnover; see also the discussion in Chapter 6 of performance under stress). Ideas and alternative plans are often needed for solving difficult and challenging problems or for removing obstacles that thwart taskwork, teamwork, and mission accomplishment. Much is known, and yet much more needs to be known, about meeting these needs when attempting to hire employees who will not only adapt successfully as newcomers to an organization but who will, over time, adapt to change and create change as well.

One can think of examples of exceptional problem solving in real or fictional life-threatening situations. Recall the NASA specialists who adapted materials available on the ill-fated, crippled, moon-bound Apollo 13 spacecraft to bring the astronauts safely back to earth. Or remember the weekly episodes of the ABC television series *MacGyver*, the ingenious troubleshooter who solved problems with everyday materials he found at hand.

Clearly, civilian organizations and the military alike seek to hire talent who can work effectively individually and in teams to solve problems that are critical to their missions. The scientific community has demonstrated without question the importance of cognitive ability, cognitive flexibility, motivation, and team coherence and coordination in solving such problems (Bell and Kozlowski, 2002; Chen et al., 2002; Ilgen et al., 2005; Salas et al., 2005). Research indicates that in many problem-solving situations where prior training or available materials are inadequate, it is often not the smartest person on the team that comes up with a solution to a problem (Mason and Watts, 2012; Woolley et al., 2010). Who are these employees and soldiers who can adapt and innovate in changing, even stressful, circumstances? What characteristics differentiate them from others?

What Is Adaptability? What Is Inventiveness?

Adaptability refers to the ability to adjust and accommodate to changing and often unpredictable physical, interpersonal, cultural, and task environments. People who are adaptable are often described as cognitively and temperamentally flexible, resilient, and hardy, actively accommodating and adjusting to uncertainty and ambiguity even under duress.

Inventiveness involves innovative thinking and the ability to produce novel ideas that are of high quality and task-appropriate,[1] especially in work settings that require practical and concrete solutions often of a mechanical nature. Inventive people are often described as ingenious, creative, original, and clever. Innovative thinking can lead to outcomes that range from everyday problem solving to transformational, paradigm-changing outcomes—all of which require novel, high-quality, task-appropriate think-

---

[1] Creativity is often defined as the ability to produce high-quality task-appropriate, novel ideas (see Sternberg, 2001). Creativity in comparison to inventiveness embraces artistic creativity, whereas inventiveness is more descriptive and useful in realistic occupations such as those found in the military.

ing. But inventiveness is more than generating ideas; it also incorporates an action orientation focused on problem solving.

Researchers in many disciplines have examined adaptability and inventiveness. Studies focusing on learning agility (the ability to learn from experience—from successes but especially from mistakes), fluid intelligence, thinking biases, intellectual engagement, domain-specific knowledge, idea generation, personality, motivation, and interests have all contributed to our understanding of inventive, adaptive behavior. The present chapter focuses specifically on idea generation and temperament (personality) variables as indicators of adaptability and inventiveness. Both fluid intelligence (especially spatial ability; see Chapter 4) and cognitive biases also play a role in inventive, adaptive behavior, but these are dealt with separately in Chapters 2 and 3, respectively. Likewise, contextual and environmental factors also play critical roles in fostering and inhibiting adaptive, inventive behavior, but these roles are outside the scope of this report, which focuses on measuring critical individual differences.

Other Relevant Constructs

Other constructs that are related to adaptability and inventiveness have unique aspects of their own worthy of discussion. Two of them, learning agility and intellectual engagement, are each compound variables that capture content from multiple constructs from different individual-differences domains. These compound constructs have been found to be successful in predicting performance in challenging situations that require new solutions, as described below.

Learning Agility

Learning agility can be defined as the willingness and ability to learn from experience of both success and failure and to apply that learning later, often under stress in new or first-time conditions (De Meuse et al., 2010). Lombardo and Eichinger (2000) described learning agility as consisting of four components: people agility, results agility, mental agility, and change agility, with each component incorporating both cognitive and noncognitive elements.[2] Similarly, and likely with more complexity, Koutstaal (2012) examined the agile mind from a multidisciplinary perspective (developmental psychology, social psychology, and neuropsychology), incorporating cognition, action, perception, and emotion. Others, such as DeRue and colleagues (2012a) argued for a narrower conceptualization of

---

[2] More recently, they have added a fifth factor—self-awareness—to the definition of learning agility (De Meuse et al., 2012).

learning agility, defining it as "speed and flexibility of learning" (DeRue et al., 2012b, p. 318). Although they state that they " . . . do not believe that learning agility is a purely cognitive process" (DeRue et al., 2012b, p. 319), several academics and practitioners have criticized their narrow definition, in that reducing the concept of learning agility to speed and flexibility of learning eliminates the most interesting and practical aspects of the concept as it applies in work and temporal contexts, such as motivation to learn, emotional regulation, and learning from prior failures and successes (see, for example, Carette and Anseel, 2012; De Meuse et al., 2012; Hezlett and Kuncel, 2012). The committee tends to agree with this latter point of view.

Typical Intellectual Engagement

Intelligence is often, perhaps typically, thought of as and measured under "maximal" performance conditions, when motivation is highest. Certainly that is often true when some form of an intelligence or achievement test is administered to employees or students. Yet intelligence is surely just as important to understand (if not more important) in day-to-day work experiences when motivation to perform is not always maximal. Goff and Ackerman (1992) introduced the concept of *typical intellectual engagement* (TIE) as an individual-differences variable that might account for differences in the expression of intelligence in everyday life. They defined TIE as a desire to engage and understand the world, an interest in a wide variety of things, a preference for thorough understanding of a topic or problem, a need to know.[3] Some have argued that TIE is little more than what is found in other personality variables such as openness to experience (e.g., Rocklin, 1994) and need for cognition (Woo et al., 2007). Facets of the Big Five factor Openness to Experience (such as intellectual efficiency, ingenuity, and curiosity) are also likely relevant to TIE. The point here is to say that TIE, Openness (the Big Five factor), and need for cognition are useful constructs that (1) encompass both cognitive and temperament (personality) variables, (2) are different from traditional intelligence measures that focus on maximal motivation of the test taker, (3) are clearly relevant to adaptability and inventiveness, and (4) warrant closer examination of their relationship to each other.

Structure of the Chapter

The remainder of this chapter consists of five sections. Section 2 is devoted to issues related to understanding and measuring outcomes that the individual-differences variable adaptability/inventiveness and its facets

---

[3] Von Stumm and colleagues (2011) refer to TIE as intellectual curiosity and "hungry mind."

should predict. The usual outcome variables, such as education, training, overall job performance, and turnover are important but inadequate. New criterion measures that are theoretically related to adaptability and inventiveness are needed. Section 3 discusses a specific cognitive ability: idea generation measured in the context of maximal motivation to perform.[4] Section 4 reviews temperament (personality) measures of behavior under typical motivational circumstances. In both Sections 3 and 4, the committee reviews the evidence indicating the relevance of measures of these constructs for predicting adaptable behavior in changing, challenging situations that require flexibility and innovative problem solving. Section 5 presents the committee's conclusion based on the research evidence, and Section 6 lists our recommendations for future research on adaptive behavior.

In short, this chapter informs readers about the importance of adaptability and inventiveness constructs in personnel selection and classification contexts. Interestingly, personnel selection and classification in the 21st century is itself a problem that requires creative and adaptable researchers and practitioners.

ADAPTABILITY/INVENTIVENESS AS AN OUTCOME VARIABLE

It is important to distinguish between adaptability and inventiveness as stable traits on which people differ and to distinguish these traits from behavioral outcomes. Pulakos and colleagues (2000), for example, examined adaptability as an outcome variable, concluding that it consists of eight components. Keeping predictors and behavioral outcomes (criteria) conceptually distinct allows one to draw empirical distinctions within testable models that ask questions such as (a) How do personality characteristics affect training success? (b) How do personality and training together predict relevant work outcomes? (c) How strong and for how long does training and past behavior predict future behavior?

The challenge of defining and measuring adaptive/creative outcomes makes validity studies challenging as well. For example, experts may disagree on whether an outcome is in fact creative; just because a person can generate a large number of solutions to a problem does not mean the solutions are any good, or conversely, just because a person generates only one solution to a problem does not mean the person (or that solution) is not adaptive or creative. Furthermore, the type of creative output might be numerous yet also constrained and specific to a particular domain of knowledge or expertise: an elegant set of computer code, a technological invention and its patent, a tricky multicultural military negotiation that is

---

[4] Spatial ability is another cognitive ability that is important in understanding and predicting adaptability and inventiveness (Kell et al., 2013; also see Chapter 4 of this report).

handled effectively, or a soul-stirring musical performance. Furthermore, sometimes the adaptive or creative nature of human behavior is not discovered within the solution but instead is reflected in reframing the problem creatively such that it, in turn, can make resulting strategies and solutions obvious and even mundane. Despite this heterogeneity, a useful conceptual framework for characterizing all individual outcomes of creativity might include *frequency/fluency* and *quality/usefulness* as reliable qualities of various adaptive and creative behaviors and output, as well as the inventiveness/novelty and radicalness/surprise of inventive or creative solutions to problems (Simonton, 2012; West and Anderson, 1996).

Social context and social networks, including teammates or coworkers, supervisors and their leadership style, and organizational climate, influence most forms of employee or soldier adaptability and creativity (Hon et al., 2014; Zhang and Zhou, 2014). Other situational factors matter as well; more complex tasks, highly stressful or emergency situations, and socially ambiguous contexts all have strong unpredictable elements to them, and unpredictability is a theme that appears to encourage and accentuate individual differences in adaptive and creative problem solving in the workplace (Pulakos et al., 2000).

To summarize, the committee emphasizes the distinction between adaptability and creative performance/outcomes and adaptability and inventiveness as individual differences that predict this type of performance. We are also sensitive to the importance of how numerous contextual factors moderate effects on empirical validity findings in this domain (Zhou and Hoever, 2014). This chapter and the entire report focus on predictors of behavioral and performance outcomes at the individual level as opposed to the team or unit level.

ADAPTABILITY/INVENTIVENESS AS AN INDIVIDUAL-DIFFERENCES COGNITIVE VARIABLE: IDEA PRODUCTION MEASURES INCREMENT VALIDITY OVER GENERAL COGNITIVE ABILITY

Carroll's (1993) reanalysis of 467 datasets identified Idea Production (he also called it Retrieval Ability) as one of eight second-stratum factors underlying general cognitive ability.[5] Idea Production usefully summarized correlations among nine first-stratum factors, including Originality/Creativity (the other eight factors were Ideational Fluency, Associational

---

[5] In this chapter, idea generation is used interchangeably with idea production. However, idea generation is often used as a descriptor of a kind of task (e.g., ideational fluency) that measures the idea production factor.

Fluency, Expressional Fluency, Word Fluency, Figural Fluency, Naming Facility, Sensitivity to Problems, and Figural Flexibility).

All the Idea Production factors and tests involve the production of ideas as opposed to recognition or comparison of them; the implication of this distinction is that idea production tasks require open-ended/recall responses, rather than multiple-choice/recognition responses. For example, listing things that are red, writing antonyms to a specified word, or listing ways that a brick can be used are all Idea Production tasks (see Box 7-1). Within this set of Idea Production factors, the Originality/Creativity factor tests are differentiated from other Idea Production tests in that "they require examinees fairly quickly to think of . . . a series of responses fitting

---

BOX 7-1
Idea Production Factor (Test Type) with Sample Items

Idea Production Factor	Sample Items (example test and typical item)
Originality/Creativity	Consequences ("what would happen if people did not have to eat?")
Ideational Fluency	Related things ("name all the red things you can think of")
Naming Facility	Picture name ("list names for a picture")
Associational Fluency	Synonyms ("list synonyms of the word *good*")
Expressional Fluency	Similes ("her eyes twinkled like ___.")
Word Fluency	First and last letter ("name words that begin with *g* and end with *t*")
Sensitivity to Problems	Improvements ("identify ways to improve the telephone")
Figural Fluency	Sketches ("add details to simple objects to make new ones")
Figural Flexibility	Match ("rearrange matchsticks to make new figures")

NOTE: Various scores can be computed from the examinee's responses.
SOURCE: Box created from research presented in Carroll (1993).

the requirements of the task . . . furthermore, . . . it is difficult and challenging to think of responses beyond the more obvious commonsense ones" (Carroll, 1993, p. 428). That is, Carroll suggested that creativity tests can be thought of as difficult fluency tests, ones requiring the rapid generation of appropriate responses but where the appropriate responses beyond the first few are nonobvious.

Evidence of Predictive Validity

This section describes several independent studies that show incremental validity of idea generation test scores over other cognitive ability measures in predicting significant real-world outcomes.

Studies on Creativity by the U.S. Army Research Institute for the Behavioral and Social Sciences

In one important study from the U.S. Army Research Institute for the Behavioral and Social Sciences, scores from the *Consequences* test were shown to be strong predictors of leadership abilities and Army officer career outcomes,[6] independent of other cognitive ability predictors (Mumford et al., 1998). The authors administered a five-item version of the test to 1,819 U.S. Army officers, along with measures of verbal reasoning and leadership expertise. Officers were asked to work on five *Consequences* problems by first reading through the description of each situation and then listing as many significant outcomes of the situations as possible in the allotted time. They were given 12 minutes to complete all five problems.

This consequences test demonstrated predictive validity with respect to several important outcomes, including career continuance, career progression, and performance at both the junior and senior levels. The main finding was that the *Consequences* test, when scored various ways and with all the scores entered in a multiple regression equation, predicted all of the outcomes, ranging from $R = .22$ (for critical incidents) to $R = .58$ (for rank). This predictive validity held up even after controlling for cognitive ability and expertise (incremental R^2 ranging from .06 to .22, with a median of .20). This finding of incremental validity stands out in the scheme of what we know about cognitive abilities measurement. Other than with personality measures, it is rare to find cognitively oriented measures providing much incremental validity over general cognitive ability in predicting broad, real-world outcomes (e.g., Humphreys, 1986; Ree and Earles, 1991).

---

[6] An important distinction relevant to military performance is that leadership is a behavior exhibited by officers and enlisted soldiers alike.

Studies on Creativity by Educational Testing Service

In a series of studies, Educational Testing Service researchers (Bennett and Rock, 1995; Frederiksen and Ward, 1978) found that idea generation (also sometimes referred to as idea production) measures, specifically ones obtained from tests of formulating hypotheses, measuring constructs, evaluating proposals, and solving methodological problems, predicted graduate school outcomes beyond what could be predicted by verbal and mathematics scores on the graduate records examination (GRE) test. The tests were originally developed by Frederiksen and Ward (1978) based on critical incident studies (Flanagan, 1954).

The tests were given to 3,586 examinees as part of an experimental section of the graduate record examinations (GRE) test. Several scores were generated from the tests, including a number score, a number of unusual responses score, a number of quality responses score, and variations on these. After they had completed a year of graduate school (a year and a half after the initial test administration), students were tested again. Significant relationships were found for a number of outcomes, including various measures of professional activities such as the number of professional activities engaged in ($r = .24$), whether they engaged in collaborative research ($r = .18$), and number of publications ($r = .18$). N's ranged from 525 to 650. Interestingly, there were no significant relationships found between these outcomes and GRE scores, suggesting that these idea production scores were related to and predictive of important school outcomes that the other standardized measures neither related to nor predicted. Bennett and Rock (1995) replicated these findings with a computer administration, albeit with a much smaller sample size. They also administered several additional tests including a *Topics* test (suggest ideas about a train journey), a *Pose-a-Question-to-a-Cardboard-Box* test (from Torrance, 1974), and two pattern-meaning items (Wallach and Kogan, 1965) that required examinees to imagine what an unfinished drawing would look like if finished. The score was simply the sum of the number of responses given by the examinee across all four items. This score was highly correlated with the formulating hypotheses test score ($r = .65$) and weakly correlated with undergraduate grade point average ($r = .27$). These findings suggest that idea generation is an ability not well reflected in standardized tests of verbal and mathematical reasoning but at the same time predictive of important outcomes. (The committee notes that some of the incremental validity of the new measures over GRE scores might be attributed to range restriction on GRE scores.)

Idea Generation Scales: Measurement Issues

Types of Scales

As noted above, Carroll (1993) summarized the variety of measures used to assess creativity in individual-differences studies. As can be seen in Box 7-1 (above), tests involve open-ended prompts, with instructions for the examinee to produce as many responses as possible within a set period of time (typically, a minute to a few minutes).

Scoring Methods

Traditionally, there are three alternative scoring approaches for idea production tests (Mumford and Gustafson, 1988):

(a) Fluency scores (number of responses/number of alternative solutions);
(b) Flexibility scores (number of shifts or variability in response categories); or
(c) Originality scores (novelty of proposed alternatives).

Existing measures have been individually hand-scored to one or all of these criteria (or even more specific criteria, e.g., Mumford et al., 1998), sometimes with the aid of rubrics but nevertheless in a labor-intensive fashion. Developments in natural language processing technology might now enable more efficient computerized scoring, resulting in operational feasibility. For example, machine scoring of essays routinely outperforms human scoring and is now commonplace in the testing industry (Shermis and Burstein, 2013).

Scoring methods: Sample studies Carroll (1993) identified fluency and originality as among the most common scoring methods in use with creativity tests. Frederiksen and Ward (1978) scored formulating hypotheses responses six ways: (a) number of hypotheses (a fluency score), (b) number of unusual (identified by fewer than 5 percent of examinees) hypotheses (an originality score), (c) number of unusual-and-high-quality hypotheses, (d) mean quality of hypotheses, (e) highest quality of any hypothesis, and (f) quality of the hypothesis marked "best." Agreement among raters for all six of these scores was fairly high (alphas ranged from .69 to .90): Coefficient alphas (Cronbach, 1951) were computed for single items, based on categorizations by two independent scorers. However, due to the brief length of the test (four items), test reliability ranged from fairly low for some of the tests on some of the scores, such as the unusual-and-high-quality score on the measuring constructs task ($r_{xx'}$ = .34), to fairly high, such as the mean quality score on that same task ($r_{xx'}$ = .88). Interestingly, although this was a high-ability sample (graduate school applicants), examinees averaged only about 2.5 responses for formulating hypotheses and solving methodological problems and slightly more for measuring constructs. The number of responses scored as unusual and as unusual-and-high-quality was only about one-third or one-quarter of these. Despite the low mean scores, the measures nevertheless correlated with outcomes, even after controlling for standardized test scores. It is possible to develop easier items, that is, items that yield higher mean scores. This could be done either through the development of easier prompts (ones that enabled more responses to be given to them) or longer response times. Presumably, with research, it would be fairly easy to develop prompts and response time windows appropriate for the enlisted applicant population.

The original scoring system for the *Consequences* test (Christensen et al., 1953; 1958) classified responses as (a) remote, (b) obvious, and (c) irrelevant, with separate scores given for remote and obvious responses. Typically, examinees generate about 4 obvious responses per item in the 2-minute time window (in Mumford et al., 1998, they were given slightly more time), and 1 or 2 remote responses (similar to the Frederiksen-Ward results for a different test). The number of remote responses was the score most likely to show high correlations with outcomes, according to the initial reports (Christensen et al., 1958). Mumford and colleagues' more recent scoring method (1998) classified responses on several dimensions (see Table 7-1). They found that scores from the first six of these eight dimensions were highly correlated, ranging from r = .64 to .92. Note that positive and negative consequences were independent of the other scores and also did not correlate with the outcomes.

The *Consequences* test has consistently shown low reliability (Gleser, 1965). Dela Rosa and colleagues (1997) suggested reverting back to something closer to the Christensen scoring approach, in which responses were scored as obvious, remote, duplicate, or irrelevant/unratable. *Obvious* responses were those that directly resulted from the situation, *remote* responses were those that referred to indirect results and differed from the material presented, duplicate responses were those restating an idea (or restating one of the given responses), and *irrelevant/unratable* responses were all others. An ideational fluency score was computed as the sum of obvious responses, and an originality score was computed as the sum of remote responses. With this scheme, Dela Rosa and colleagues (1997) were able to attain reasonable reliabilities: for three raters, $r_{xx'}$ = .82 to .92 for ideational fluency, and $r_{xx'}$ = .86 to .94 for originality. Milan

TABLE 7-1 Scoring Schemes Applied by Mumford and Colleagues (1998)

Criterion	Scale	Description
1. Quality	5-point-scale	How coherent, meaningful, and logical are the consequences with respect to the question being asked?
2. Originality	5-point-scale	To what degree are the consequences novel and imaginative? To what extent do they differ from the material presented or state more than what is obviously apparent from the problem? This also refers to the degree to which obvious consequences are presented with unique or unusual implications.
3. Time Frame	5-point-scale	How realistic and pragmatic are the consequences and would they occur in the real world?
4. Realism	5-point-scale	To what extent do the consequences focus on long-term implications as opposed to short-term or immediate concerns?
5. Complexity	5-point-scale	The degree to which the consequences contain multiple elements and describe the interrelations among those elements.
6. Use of General Principles	5-point-scale	To what degree are there principles, laws, procedures, etc. underlying the consequences.
7. Positive Consequences	(yes/no)	Refers to the presence or addition of something.
8. Negative Consequences	(yes/no)	Refers to the absence or diminishment of something.

SOURCE: Mumford, M.D., A. Michelle, M.S. Connelly, S.J. Zaccaro, and J.F. Johnson. (1998). Domain-based scoring in divergent-thinking tests: Validation evidence in an occupational sample. *Creativity Research Journal, 11*(2):155. Reproduced by permission of Taylor & Francis, Ltd., http://www.tandfonline.com. Scale column added by committee. Criterion numbers assigned by committee and description slightly edited.

and colleagues (2002) found that the ideational fluency and originality scores were fairly independent ($r = .10$).

Scoring approaches: Summary findings Box 7-2 lists five idea generation studies, identifying the test and the criteria.

> **BOX 7-2**
> **Five Important Idea Generation Studies**
>
> Bennett and Rock (1995)   Test: formulating hypotheses;
>   Criteria: GRE scores and GPA
>
> Bennett and Rock (1998)   Test: generating explanations;
>   Criteria: GRE scores and GPA
>
> Frederiksen and Ward   Test: formulating hypotheses;
> (1978)   Criteria; scores on several cognitive tests
>   (e.g., verbal, quantitative)
>
> Hoover and Feldhusen   Test: formulating hypotheses;
> (1990)   Criteria: scores on several ognitive tests
>   (e.g., abstract reasoning, verbal, quantitative,
>   speed)
>
> Mumford et al. (1998)   Test: consequences;
>   Criteria: several different indicators of
>   organizational leadership
>
> NOTE: Based on studies found during the committee's literature review. GPA = grade point average, GRE = graduate records examination.

ADAPTABILITY/INVENTIVENESS AS AN INDIVIDUAL-DIFFERENCES NONCOGNITIVE VARIABLE

The previous section described in depth the evidence that combining a measure of general cognitive ability with a measure of idea generation increases the accuracy of predicting important outcomes. The same is true for adding personality variables to the equation: accuracy of predicting important outcomes increases. When the validity of specific cognitive abilities (e.g., divergent thinking and spatial ability) and personality variables (e.g., achievement motivation, dominance, and creative personality) are included in a predictor battery, correlations for predicting innovative contributions are high (observed $r = .53$; $\rho = .58$) (Hough and Dilchert, 2007).

To draw on the literature on personality as it pertains to adaptability and inventiveness, the committee first briefly reviews the nature and structure of personality constructs. After that, the discussion turns to an examination of empirically supported relationships between personality variables and innovative contributions/outcomes. The personality variables that are implicated in the sections that follow are suggestive of the complement of individual differences required to understand individual differences in adaptability and inventiveness.

Structure of Personality

The Five-Factor Model[7] of personality variables is often used to organize the myriad of personality variables that psychologists use to study personality. It is a hierarchical model with five broad factors: Emotional Stability, Extraversion, Conscientiousness, Agreeableness, and Openness to Experience.

There are also correlated subdimensions, often referred to as facets, associated with each of the five factors. Although there is agreement within the personality community on the existence of facets, there is less agreement on their specific identity. Different investigators tend to identify different facets. There have been several attempts to try to characterize commonalities across investigators. One by John and colleagues (2008), which is represented in Table 7-2, examines the overlap between facets measured by the NEO Personality Index R (NEO-PI-R, Costa and McCrae, 1992),[8] by a lexical-based facets instrument (Saucier and Ostendorf, 1999), and by the California Psychological Inventory (Soto and John, 2009). It appears from this analysis that approximately two or three facets per factor overlap across the three schemes, but others are unique. The solution developed by Drasgow and colleagues (2012) represents an amalgamation or summary of several solutions. They began with the findings of Saucier and Ostendorf (1999) as their first input. They then considered International Personality Item Pool data from the Oregon-Eugene-Springfield community samples ($N = 727$ adult volunteers) on seven personality inventories: the NEO-PI-R, the 16PF (16 Personality Factors), the California Psychological Inventory, the Manchester Personality Questionnaire, the Jackson Personality Inventory, the Hogan Personality Inventory, and the Abridged Big Five-Dimensional Circumplex. They analyzed data from the various inventories, one Big Five scale at a time. Then, using exploratory factor analysis, they identified 22 stable facets that fall under the broader Big Five factors. The Big Five factors along with the 22 facets appear in Table 7-2. Their solution represents a comprehensive mapping of facets of various inventories or researchers that embrace the Five-Factor Model of personality.

Other approaches to examining the structure of personality include the

---

[7] "Five-Factor Model" is an alternative name for the Big Five personality factor model. Both names are used interchangeably in this chapter.
[8] NEO-PI-R measures neuroticism, extraversion, openness to experience, agreeableness, and conscientiousness, as well as subordinate dimensions.

TABLE 7-2 Defining Facets for the Big Five Factors: Four Solutions

Lexical Facets (18) (Saucier and Ostendorf, 1999)	NEO-PI-R Facets (30) (Costa and McCrae, 1992)	CPI–Big Five Facets (16) (Soto and John, 2009)	TAPAS (22) (Drasgow et al., 2012)
Extraversion Facets			
Sociability	Gregariousness	Gregariousness	Dominance
Assertiveness	Assertiveness	Assertiveness/Leadership	Activity
Activity/Adventurousness	Activity	Social Confidence vs. Anxiety	Sociability
Unrestraint	Excitement-Seeking		Attention Seeking
	Positive Emotions		
	Warmth		
Agreeableness Facets			
Warmth/Affection	Modesty	Modesty vs. Narcissism	Cooperation
Modesty/Humility	Trust	Trust vs. Suspicion	Consideration
Generosity	Tender-Mindedness	Empathy/Sympathy	Selflessness
Gentleness	Compliance	Altruism	
	Straightforwardness		
Conscientiousness Facets			
Orderliness	Order	Orderliness	Order
Industriousness	Achievement Striving	Industriousness	Achievement
Reliability	Dutifulness	Self-Discipline	Self-Control
Decisiveness	Self-Discipline		Responsibility
	Competence		Non-Delinquency
	Deliberation		Virtue
Neuroticism Facets			
Insecurity	Anxiety	Anxiety	Optimism
Emotionality	Anger/Hostility	Irritability	Adjustment
Irritability	Depression	Depression	Even-Tempered
	Self-Consciousness	Rumination-Compulsiveness	
	Vulnerability		
	Impulsiveness		
Openness Facets			
Intellect	Ideas	Intellectualism	Aesthetics
Imagination/Creativity	Aesthetics	Idealism	Intellectual Efficiency
Perceptiveness	Fantasy	Adventurousness	Tolerance
	Actions		Ingenuity
	Feelings		Depth
	Values		Curiosity

SOURCE: John, O.P., L.P. Naumann, and C.J. Soto. (2008). Paradigm shift to the integrative Big Five trait taxonomy: History, measurement, and conceptual issues. In O.P. John, R.W. Robins, and L.A. Pervin, Eds., *Handbook of Personality: Theory and Research* (p. 126). New York: Guilford Press. Reproduced by permission conveyed through Copyright Clearance Center.

HEXACO model[9] (Ashton and Lee, 2001; Ashton et al., 2004a, 2004b) and Hough's nomological web clustering approach (Hough and Ones, 2001; Hough et al., 2015; Oswald et al., 2013). The HEXACO model organizes facet-level variables using a six-factor, hierarchical circumplex model in which relationships between personality characteristics are envisioned as a circle with two factors, one on each axis. Hough's nomological web clustering approach organizes personality variables into clusters that demonstrate very high construct validity (including convergent and discriminant validity) based on correlational evidence between personality variables, factor and component analysis, expert judgments, criterion-related validities between personality variables and outcome variables, and indices of subgroups of people (e.g., ethnic groups, men and women). Both approaches are nonhierarchical; that is, they acknowledge the reality of complex relationships between personality variables wherein facets in one factor correlate with facets in other factors more highly than they do with facets in the factor to which they supposedly belong, a phenomenon that should not occur if the model is envisioned as hierarchical.

The point is that personality variables defined and measured more narrowly than at the broad level of the Five-Factor Model are likely to yield stronger correlations with outcome variables measuring adaptability and inventiveness. In short, facet-level personality variables, which may or may not be later combined, warrant further research.

Evidence of Validity of Personality Variables Predicting Adaptive/Innovative Outcomes

Numerous meta-analyses have used the Five-Factor Model to summarize criterion-related validities of personality variables for predicting work-related outcomes, including adaptive/innovative/creative contributions. Table 7-3 organizes meta-analytic and single-study correlational evidence of the relationships between personality variables and adaptive/innovative outcomes (criteria) using the Five-Factor Model.[10]

---

[9] The HEXACO personality inventory assesses honesty-humility, emotionality, extraversion, agreeableness, contentiousness, and openness to experience.

[10] Validity studies of personality variables are sometimes criticized because researchers involved in some of the studies have financial interests in one or more of the personality measures. It is the experience of the committee members who have developed personality measures (the majority of which do not have financial interest in any personality measure) that the validities in Table 7-3 are representative of findings that they have observed. The committee also points out that this criticism is not typically directed at criterion-related validity studies involving cognitive abilities that were undertaken by developers of cognitive ability tests who had financial interests in those cognitive ability tests.

These data indicate the following:

- Some personality variables are related to adaptive/inventive outcomes, whereas other personality variables are not. For example, for the Five-Factor Model:
  o Emotional Stability predicts some types of adaptive/inventive outcomes.
  o Some facets of Conscientiousness predict adaptive/inventive outcomes.
  o Some facets of Extraversion predict adaptive/inventive outcomes.
  o Some facets of Openness to Experience predict inventive outcomes but do not appear to predict either proactive or reactive forms of adaptive behavior.
  o Composites (compound variables) that comprise relevant personality variables predict adaptive/inventive outcomes better than any personality variable used individually.
- In several cases, a facet-level variable is a stronger predictor of adaptive/inventive outcomes than its umbrella Big Five construct. For example:
  o The Big Five construct Extraversion does not appear to predict adaptive/inventive outcomes but two of its facets, Dominance and Activity/Energy, do predict adaptive/inventive outcomes, whereas Sociability, another Extraversion facet, does not.
  o The Big Five construct Conscientiousness, does not appear to predict adaptive/inventive outcomes. But, again, one of its facets, Achievement, does predict adaptive/inventive outcomes, whereas Deliberation/Cautiousness, another Conscientiousness facet, does not.
- The type of job and the type of adaptive outcome moderate the relationship between personality and adaptive outcomes. In particular, adaptive performance outcomes may be proactive or reactive in nature, where proactive forms deal with people identifying a need to change the environment when it is relatively constant and reactive forms deal with people needing to adapt whenever the environment changes. Some findings are as follows:
  o Achievement (a facet of Conscientiousness) predicts proactive forms of adaptive performance for managers ($\rho = .28$) better than reactive forms of adaptive performance ($\rho = .20$).
  o Similarly, for nonmanagerial employees, Achievement (facet of Conscientiousness) predicts proactive forms of adaptive performance ($\rho = .14$) better than reactive forms of adaptive performance ($\rho = .11$), although clearly Achievement is a better predictor of

proactive forms of adaptive performance for managers than it is for nonmanagerial employees.
- o On the other hand, Emotional Stability predicts reactive forms of adaptive performance for managers ($\rho = .25$) better than proactive forms of adaptive performance ($\rho = .15$).
- o Similarly, for nonmanagerial employees, Emotional Stability predicts reactive forms of adaptive performance ($\rho = .18$) better than proactive forms of adaptive performance ($\rho = .11$).
- o In both cases, personality variables predict adaptive behavior more strongly for managers than for nonmanagers.
- The determinants of adaptive/inventive outcomes are complex. Advances in scientific understanding of the role of personality in determining these outcomes will likely come from understanding facets and other personality constructs that are narrower than broad factors.

The majority of the studies in the meta-analyses and single studies listed in Table 7-3 are concurrent (rather than predictive) validity studies. They are indicative of the relationships between personality variables and adaptive/innovative outcomes and are thus instructive for identifying measures (especially facet-level measures) of personality characteristics that are most likely to be predictive of adaptive/innovative outcomes. The values of the better personality predictors in Table 7-3 are in the .20s even .30s. Given that personality variables and cognitive ability variables are only minimally correlated or uncorrelated (Ackerman and Heggestad, 1997; Judge et al., 1999; McHenry et al., 1990), the incremental validity of predicting adaptive/innovative outcomes is likely significant. The Tailored Adaptive Personality Assessment System (TAPAS), the personality inventory the military is using and continues to evaluate, includes facet-level measures such as Ingenuity, Curiosity, and Intellectual Efficiency (Stark et al., 2014). Table 7-3 indicates these scales likely measure important variance relevant to adaptability and inventiveness. Their merit for predicting adaptive/inventive outcomes needs to be researched.

Nonetheless, given the extent to which coaching and intentional distortion in high-stakes employment settings occurs, validities may be lower. This issue and advances in personality test development that address this issue are discussed below.

Measurement Issues: Personality

Use of self-report personality measures to select among applicants for a desirable job or school is frequently criticized because respondents can lie about themselves on positive traits, thus improving their scores on such

TABLE 7-3 Criterion-Related Validities of Big Five, Facet-Level, Compound, and Other Personality Variables

Personality Variable	Adaptive/Innovative Outcomes	
	Observed Validity*	Corrected Validity*
Big Five and Facets		
Emotional Stability	$r = -.07$; $k = 128$; artists vs. non-artists (Feist, 1998)	$\rho = -.03$; $k = 4$; $N = 1,332$; lab (Harrison et al., 2006)
	$r = -.05$; $k = 8$; $N = 442$ (Hough, 1992)	$\rho = .04$; $k = 3$; $N = 448$; field (Harrison et al., 2006)
	$r = .02$; $k = 66$; creative scientists (Feist, 1998)	$\rho = .15$; $k = 17$; $N = 1,823$; managers; proactive forms of adaptive performance (Huang et al., 2014)
	$r = .09$; $k = 17$; $N = 1,823$; managers; proactive forms of adaptive performance (Huang et al., 2014)	$\rho = .11$; $k = 48$; $N = 5,270$; employees; proactive forms of adaptive performance (Huang et al., 2014)
	$r = .08$; $k = 48$; $N = 5,270$; employees; proactive forms of adaptive performance (Huang et al., 2014)	$\rho = .25$; $k = 18$; $N = 1,864$; managers; reactive forms of adaptive performance (Huang et al., 2014)
	$r = .16$; $k = 18$; $N = 1,864$; managers; reactive forms of adaptive performance (Huang et al., 2014)	$\rho = .18$; $k = 51$; $N = 5,450$; employees; reactive forms of adaptive performance (Huang et al., 2014)
	$r = .13$; $k = 51$; $N = 5,450$; employees; reactive forms of adaptive performance (Huang et al., 2014)	
	$r = .18$; $N \sim 330$ (Pulakos et al., 2002)	
Extraversion	$r = .14$; $k = 135$; creative scientists (Feist, 1998)	$\rho = .04$; $k = 3$; $N = 448$; field (Harrison et al., 2006)
	$r = .08$; $k = 148$; artists vs. non-artists (Feist, 1998)	$\rho = .03$; $k = 4$; $N = 1,332$; lab (Harrison et al., 2006)
Facet: Dominance	$r = .21$; $k = 11$; $N = 550$ (Hough, 1992)	
	$r = .19$; $k = 42$; creative scientists (Feist, 1998)	
	$r = .08$; $k = 42$; artists vs. non-artists (Feist, 1998)	

continued

TABLE 7-3 Continued

Personality Variable	Adaptive/Innovative Outcomes	
	Observed Validity*	Corrected Validity*
Facet: Sociability	$r = -.25$; $k = 2$; $N = 116$ (Hough, 1992)	$\rho = .05$; $k = 17$; $N = 1,823$; managers; proactive forms of adaptive performance (Huang et al., 2014)
	$r = .07$; $k = 23$; creative scientists (Feist, 1998)	
	$r = .01$; $k = 35$; artists vs. non-artists (Feist, 1998)	$\rho = .01$; $k = 48$; $N = 5,270$; employees; proactive forms of adaptive performance (Huang et al., 2014)
	$r = .04$; $k = 17$; $N = 1,823$; managers; proactive forms of adaptive performance (Huang et al., 2014)	$\rho = .02$; $k = 18$; $N = 1,864$; managers; reactive forms of adaptive performance (Huang et al., 2014)
	$r = .01$; $k = 48$; $N = 5,270$; employees; proactive forms of adaptive performance (Huang et al., 2014)	$\rho = -.01$; $k = 51$; $N = 5,450$; employees; reactive forms of adaptive performance (Huang et al., 2014)
	$r = .01$; $k = 18$; $N = 1,864$; managers; reactive forms of adaptive performance (Huang et al., 2014)	
	$r = .00$; $k = 51$; $N = 5,450$; employees; reactive forms of adaptive performance (Huang et al., 2014)	
	Negative; $N = 225$ (Weiss, 1981)	
Facet: Activity/ Energy	Positive (Weiss, 1981)	
Conscientiousness	$r = .07$; $k = 48$; creative scientists (Feist, 1998)	$\rho = .13$; $k = 3$; $N = 707$; lab (Harrison et al., 2006)
	$r = -.29$; $k = 52$; artists vs. non-artists (Feist, 1998)	$\rho = .00$; $k = 3$; $N = ?$ (Eder and Sawyer, 2007)
		$\rho = -.06$; $k = 5$; $N = 946$; field (Harrison et al., 2006)

TABLE 7-3 Continued

Personality Variable	Adaptive/Innovative Outcomes	
	Observed Validity*	Corrected Validity*
Facet: Dependability	$r = -.07$; $k = 5$; $N = 268$ (Hough, 1992) $r = .07$; $k = 17$; $N = 1,823$; managers; proactive forms of adaptive performance (Huang et al., 2014) $r = .05$; $k = 48$; $N = 5,270$; employees; proactive forms of adaptive performance (Huang et al., 2014) $r = .05$; $k = 18$; $N = 1,864$; managers; reactive forms of adaptive performance (Huang et al., 2014) $r = .08$; $k = 51$; $N = 5,450$; employees; reactive forms of adaptive performance (Huang et al., 2014) Negative (Welsh, 1975)	$\rho = .11$; $k = 17$; $N = 1,823$; managers; proactive forms of adaptive performance (Huang et al., 2014) $\rho = .07$; $k = 48$; $N = 5,270$; employees; proactive forms of adaptive performance (Huang et al., 2014) $\rho = .08$; $k = 18$; $N = 1,864$; managers; reactive forms of adaptive performance (Huang et al., 2014) $\rho = .10$; $k = 51$; $N = 5,450$; employees; reactive forms of adaptive performance (Huang et al., 2014)

continued

TABLE 7-3 Continued

Personality Variable	Adaptive/Innovative Outcomes	
	Observed Validity*	Corrected Validity*
Facet: Achievement	$r = .14$; $k = 2$; $N = 116$ (Hough, 1992) $r = .18$; $k = 17$; $N = 1,823$; managers; proactive forms of adaptive performance (Huang et al., 2014) $r = .09$; $k = 48$; $N = 5,270$; employees; proactive forms of adaptive performance (Huang et al., 2014) $r = .12$; $k = 18$; $N = 1,864$; managers; reactive forms of adaptive performance (Huang et al., 2014) $r = .08$; $k = 51$; $N = 5,450$; employees; reactive forms of adaptive performance (Huang et al., 2014) $r = .31$; $N \sim 330$ (Pulakos et al., 2002) Positive (Amabile et al., 1994)	$\rho = .28$; $k = 17$; $N = 1,823$; managers; proactive forms of adaptive performance (Huang et al., 2014) $\rho = .14$; $k = 48$; $N = 5,270$; employees; proactive forms of adaptive performance (Huang et al., 2014) $\rho = .20$; $k = 18$; $N = 1,864$; managers; reactive forms of adaptive performance (Huang et al., 2014) $\rho = .11$; $k = 51$; $N = 5,450$; employees; reactive forms of adaptive performance (Huang et al., 2014)
Facet: Deliberation/ Cautiousness	Negative (Welsh, 1975) Negative (King, 1990)	

TABLE 7-3 Continued

Personality Variable	Adaptive/Innovative Outcomes	
	Observed Validity*	Corrected Validity*
Agreeableness	$r = -.29$; $k = 3$; $N = 174$ (Hough, 1992) $r = -.10$; $k = 63$; artists vs. non-artists (Feist, 1998) $r = -.03$; $k = 64$; creative scientists (Feist, 1998) $r = .09$; $k = 17$; $N = 1,823$; managers; proactive forms of adaptive performance (Huang et al., 2014) $r = .04$; $k = 48$; $N = 5,270$; employees; proactive forms of adaptive performance (Huang et al., 2014) $r = .10$; $k = 18$; $N = 1,864$; managers; reactive forms of adaptive performance (Huang et al., 2014) $r = .07$; $k = 51$; $N = 5,450$; employees; reactive forms of adaptive performance (Huang et al., 2014)	$\rho = -.04$; $k = 3$; $N = 448$; field (Harrison et al., 2006) $\rho = .08$; $k = 3$; $N = 707$; lab (Harrison et al., 2006) $\rho = .11$; $k = 17$; $N = 1,823$; managers; proactive forms of adaptive performance (Huang et al., 2014) $\rho = .04$; $k = 48$; $N = 5,270$; employees; proactive forms of adaptive performance (Huang et al., 2014) $\rho = .12$; $k = 18$; $N = 1,864$; managers; reactive forms of adaptive performance (Huang et al., 2014) $\rho = .07$; $k = 51$; $N = 5,450$; employees; reactive forms of adaptive performance (Huang et al., 2014)
Openness to Experience	$r = .21$; $k = 93$; artists vs. non-artists (Feist, 1998) $r = .18$; $k = 52$; creative scientists (Feist, 1998) $r = .06$; $k = 17$; $N = 1,823$; managers; proactive forms of adaptive performance (Huang et al., 2014) $r = .03$; $k = 48$; $N = 5,270$; employees; proactive forms of adaptive performance (Huang et al., 2014) $r = .04$; $k = 18$; $N = 1,864$; managers; reactive forms of adaptive performance (Huang et al., 2014) $r = .01$; $k = 51$; $N = 5,450$; employees; reactive forms of adaptive performance (Huang et al., 2014)	$\rho = .33$; $k = 3$; $N = 707$; lab (Harrison et al., 2006) $\rho = .29$; $k = 4$; $N = 597$; field (Harrison et al., 2006) $\rho = .17$; $k = 7$; $N = ?$ (Eder and Sawyer, 2007) $\rho = .09$; $k = 17$; $N = 1,823$; managers; proactive forms of adaptive performance (Huang et al., 2014) $\rho = .03$; $k = 48$; $N = 5,270$; employees; proactive forms of adaptive performance (Huang et al., 2014) $\rho = .07$; $k = 18$; $N = 1,864$; managers; reactive forms of adaptive performance (Huang et al., 2014) $\rho = .01$; $k = 51$; $N = 5,450$; employees; reactive forms of adaptive performance (Huang et al., 2014)

continued

TABLE 7-3 Continued

Personality Variable	Adaptive/Innovative Outcomes	
	Observed Validity*	Corrected Validity*
Compound Variables		
Creative Personality		$\rho = .17$; $k = 6$; $N = ?$ (Eder and Sawyer, 2007)
		$\rho = .17$; $k = 5$; $N = 1,031$ (Harrison et al., 2006)
		$\rho = .17$ business creation; $k = 15$; $N = 4,620$ (Rauch and Frese, 2007)
		$\rho = .27$ entrepreneurial success; $k = 7$; $N = 800$ (Rauch and Frese, 2007)
Specific Compound Scales:		
Creative Personality Scales[1]	$r = .26$; $k = 15$; $N = 1,086$ (Hough and Dilchert, 2007)	$\rho = .37$; $k = 15$; $N = 1,086$ (Hough and Dilchert, 2007)
Achievement via Independence[2]	$r = .26$; $N = 1,028$ (Gough, 1992)	
Independence[3]	$r = .27$; $N = 1,028$ (Gough, 1992)	
Intellectual Efficiency[2]	$r = .25$; $N = 1,028$ (Gough, 1992)	
Flexibility[2]	$r = .18$; $N = 1,028$ (Gough, 1992)	
Tolerance[2]	$r = .22$; $N = 1,028$ (Gough, 1992)	
Cognitive Flexibility[4]	$r = .17$; $N = 1,028$ (Gough, 1992)	
Complexity/ Simplicity[5]	$r = .25$; $N = 1,028$ (Gough, 1992)	
Inquiringness[4]	$r = .18$; $N = 1,028$ (Gough, 1992)	

*The statistic r (in comparison with r^2) is a direct measure of predictive efficiency (Brogden, 1946; Campbell, 1976).
[1] Affective Check List (Gough, 1979; Gough and Heilbrun, 1983).
[2] California Psychological Inventory (Gough, 1996).
[3] Barron Independence Scale (Barron, 1953b).
[4] Differential Reaction Schedule (Gough, 1962).
[5] Barron Complexity/Simplicity Scale (Barron, 1953a).

tests, and studies show that test takers are certainly able to improve their scores on self-report personality tests.[11] The concern, then, is if all applicants describe themselves in ways that increase their chances of getting the job or school offer, the resulting scores will have little to no variance and thus will no longer predict outcomes of interest. Issues involved in faking, intentional distortion, and coaching are complex; simplistic claims such as "faking doesn't matter" or "faking renders personality tests useless" are unwarranted (Hough and Connelly, 2012; Hough and Johnson, 2013; Oswald and Hough, 2011).

Personality test items are easy to fake when presented as statements tied to a rating scale response format (e.g., "indicate level of agreement with the statement 'I work hard' on a scale ranging from 'strongly agree' to 'strongly disagree'"). An examinee simply "strongly agrees" to a socially desirable statement (and "strongly disagrees" with an undesirable one). Thus, in the high stakes use of personality testing, the forced choice is a popular alternative response format because it makes it more difficult to fake. The forced-choice method presents two or more statements and asks examinees to select the one that "best describes" them and/or the one that least describes them. A limitation of forced-choice methods is that they typically yield ipsative data, which means that traits are constrained to be negatively correlated on average (by approximately $-1 / [d - 1]$, where d is the number of dimensions being compared). One way to circumvent this problem would be to test a large number of dimensions with an expectation that many of them would not be used in employment screening.

More recently, there have been developments in Item Response Theory (IRT) methods to analyze forced-choice responses that are designed to yield normative rather than ipsative data. This means that there are no constraints on the correlations between dimensions. One of these is Stark's multidimensional unfolding pairwise preference model (Stark 2002; Stark et al., 2005), which assumes that (a) personality item responses can be modeled with an unfolding model (e.g., one can fail to endorse a statement either because one is too low or too high on the trait being measured) and (b) when evaluating a pair of statements in forced-choice presentation, the examinee chooses the statement that is "closer" to him or her, which is possible because the statement can be located on the same trait continuum as the examinee. This model is implemented in the Department of Defense's application of the TAPAS (Drasgow et al., 2012).

Another IRT model for scoring forced choice is one developed by Brown and Maydeu-Olivares (2011; 2013). A strength of this model is that

---

[11] See Hough and Connelly (2012) for an indepth review of intentional distortion on self-report personality inventories.

one can apply it to existing preference data. This model has been implemented in a commercial personality assessment.

A recent meta-analysis by Salgado and Tauriz (2014) compared forced-choice methods that are scored to minimize the ipsative data constraints (which they called "quasi-ipsative") to pairs of single statements from different traits that are compared with one another and then scored (which they called "normative forced choice"). After collapsing across academic and occupational criteria, and after correcting for psychometric artifacts (measure unreliability and range restriction), the quasi-ipsative forced-choice scores for conscientiousness showed a much higher average criterion-related validity coefficient across studies (i.e., $\rho = .40$; $k = 44$) compared with normative forced-choice scoring of the same trait (i.e., $\rho = .16$; $k = 88$). It is worth noting that the correlations for quasi-ipsative forced-choice measures are higher than the correlations found in other personality meta-analyses, which have been primarily driven by results for single-statement normatively scored measures. Studies involving some of the newer IRT modeling for ipsative measures were not included in the meta-analysis, but the results for these newer approaches are likely to be more similar to the results obtained for quasi-ipsative forced-choice measures than to results for normative measures. It is important to note that these meta-analytic results for validity are averages that were associated with a vast amount of heterogeneity that remains to be explored in additional large-sample studies comparing formats in organizational samples.

Turning to the issue of the amount of faking on personality measures, the evidence seems consistent that forced-choice measures reduce score inflation (e.g., Nguyen and McDaniel, 2000). Recent research with the TAPAS (a forced-choice format) comparing scores obtained in applicant (high-stakes testing) and incumbent (low stakes, low motivation to inflate) settings in the military indicates that score inflation is minimal, about 0.15 standard deviation (Stark et al., 2014), a value much lower than found with non-forced-choice formats (Viswesvaran and Ones, 1999). Future research could be conducted to determine what aspects of forced choice are most useful for improving measurement quality and to assess the importance of various item and scoring features, such as the ideal point response process or the addition of personality items worded at a moderate level (Oswald and Schell, 2010).

RESEARCH RECOMMENDATION

The U.S. Army Research Institute for the Behavioral and Social Sciences should support research to understand constructs and assessment

methods in the domains of adaptability/inventiveness and adaptive performance, including but not limited to the following topics:

A. Compare alternative approaches to the measurement and scoring of idea generation as a cognitive measure of adaptability/inventiveness.
B. Use existing literature, theory, and empirical research to identify and develop narrow personality measures as candidates for predicting adaptive performance.
C. Develop a range of measures of relevant work criteria that reflect adaptive performance in research studies.
D. Examine the use of these personality and idea generation measures in predicting the above adaptive performance criteria.

REFERENCES

Ackerman, P.L., and E.D. Heggestad. (1997). Intelligence, personality, and interests: Evidence for overlapping traits. *Psychological Bulletin, 121*(2):219–245.

Amabile, T.M., K.G. Hill, B.A. Hennessey, and E.M. Tighe. (1994). The Work Preference Inventory: Assessing intrinsic and extrinsic motivational orientations. *Journal of Personality and Social Psychology, 66*(5):950–967.

Ashton, M.C., and K. Lee. (2001). A theoretical basis for the major dimensions of personality. *European Journal of Personality, 15*(5):327–353.

Ashton, M.C., K. Lee, and G.R. Goldberg. (2004a). A hierarchical analysis of 1,071 English personality-descriptive adjectives. *Journal of Personality and Social Psychology, 87*(5):707–721.

Ashton, M.C., K. Lee, M. Perugini, P. Szarota, R.E. De Vries, L. Di Blas, and B. De Raad, B. (2004b). A six-factor structure of personality-descriptive adjectives: Solutions from psycholexical studies in seven languages. *Journal of Personality and Social Psychology, 86*(2):356–366.

Barron, F. (1953a). Complexity-simplicity as a personality dimension. *Journal of Abnormal and Social Psychology, 48*(2):163–172.

Barron, F. (1953b). Some personality correlates of independence of judgment. *Journal of Personality, 21*(3):287–297.

Bell, B.S., and S.W.J. Kozlowski. (2002). Goal orientation and ability: Interactive effects on self-efficacy, performance, and knowledge. *Journal of Applied Psychology, 87*(3):497–505.

Bennett, R.E., and D.A. Rock. (1995). Generalizability, validity, and examinee perceptions of a computer-delivered formulating hypotheses test. *Journal of Educational Measurement, 32*(1):19–36.

Bennett, R.E., and D.A. Rock. (1998). *Examining the Validity of a Computer-Based Generating-Explanations Test in an Operational Setting* (Research Report No. RR-97–18). Princeton, NJ: Educational Testing Service.

Brogden, H.E. (1946). On the interpretation of the correlation coefficient as a measure of predictive efficiency. *Journal of Educational Psychology, 37*(2):65–76.

Brown, A., and A. Maydeu-Olivares. (2011). Item response modeling of forced-choice questionnaires. *Educational and Psychological Measurement, 71*(3):460–502.

Brown, A., and A. Maydeu-Olivares. (2013). How IRT can solve problems of ipsative data in forced-choice questionnaires. *Psychological Methods, 18*(1):36–52.

Campbell, J.P. (1976). Psychometric theory. In M.D. Dunnette, Ed., *Handbook of Industrial and Organizational Psychology* (pp. 185–222). Chicago, IL: Rand McNally.
Carette, B., and F. Anseel. (2012). Epistemic motivation is what gets the learner started. *Industrial and Organizational Psychology*, 5(3):306–308.
Carroll, J.B. (1993). *Human Cognitive Abilities: A Survey of Factor-Analytic Studies*. New York: Cambridge University Press.
Chen, G., S.S. Webber, P.D. Bliese, J.E. Mathieu, S.C. Payne, D.H. Born, and S.J. Zaccaro. (2002). Simultaneous examination of the antecedents and consequences of efficacy beliefs at multiple levels of analysis. *Human Performance*, 15(4):381–409.
Christensen, P.R., P.R. Merrifield, and J.P. Guilford. (1953). *Consequences Form A-1*. Beverly Hills, CA: Sheridan Supply.
Christensen, P.R., P.R. Merrifield, and J.P. Guilford. (1958). *Consequences: Manual for Administration, Scoring, and Interpretation*. Beverly Hills, CA: Sheridan Supply.
Cronbach, L.J. (1951). Coefficient alpha and the internal structure of tests. *Psychometrika*, 16(3):297–334.
Costa, P.T., Jr., and R.R. McCrae. (1992). *Revised NEO Personality Inventory (NEO-PI-R) and NEO Five-Factor Inventory (NEO-FFI) Professional Manual*. Odessa, FL: Psychological Assessment Resources.
Dela Rosa, M.R., D.J. Knapp, B.D. Katz, and S.C. Payne. (1997). *Scoring System Improvements to Three Leadership Predictors* (Technical Report #1070). Alexandria, VA: U.S. Army Research Institute for the Behavioral and Social Sciences.
De Meuse, K.P., G. Dai, and G.S. Hallenbeck. (2010). Learning agility: A construct whose time has come. *Consulting Psychology Journal: Practice and Research*, 62(2):119–130.
De Meuse, K.P., G. Dai, V.V. Swisher, R.W. Eichinger, and M.M. Lombardo. (2012). Leadership development: Exploring clarifying, and expanding our understanding of learning agility. *Industrial and Organizational Psychology*, 5(3):280–315.
DeRue, D.S., S.J. Ashford, and C.G. Myers. (2012a). Learning agility: In search of conceptual clarity and theoretical grounding (focal article). *Industrial and Organizational Psychology*, 5(3):258–279.
DeRue, D.S., S.J. Ashford, and C.G. Myers. (2012b). Learning agility: In search of conceptual clarity and theoretical grounding (response to commentaries). *Industrial and Organizational Psychology*, 5(3):316–322.
Drasgow, F., S. Stark, O.S. Chernyshenko, C.D. Nye, C.L. Hulin, and L.A. White. (2012). *Development of the Tailored Adaptive Personality Assessment System (TAPAS) to Support Army Personnel Selection and Classification Decisions*. Urbana, IL: Drasgow Consulting Group.
Eder, P., and J.E. Sawyer. (2007). *A Meta-Analytic Examination of Employee Creativity*. Poster presented at 22nd Annual Conference for the Society for Industrial and Organizational Psychology, New York City, NY.
Feist, G.J. (1998). A meta-analysis of personality in scientific and artistic creativity. *Personality and Social Psychology Review*, 2(4):290–309.
Flanagan, J.C. (1954). The critical incident technique. *Psychological Bulletin*, 51(4):327–358.
Frederiksen, N., and W.C. Ward. (1978). Measures for the study of creativity in scientific problem-solving. *Applied Psychological Measurement*, 2(1):1–24.
Gleser, G. (1965). Review of consequences test. In O.K. Buros, Ed., *The Sixth Mental Measurements Yearbook*. Lincoln, NE: Buros Institute of Mental Measurements.
Goff, M., and P.L. Ackerman. (1992). Personality-intelligence relations: Assessment of typical intellectual engagement. *Journal of Educational Psychology*, 84(4):537–552.
Gough, H.G. (1962). Imagination—undeveloped resource. In S.J. Parnes and H.F. Harding, Eds., *A Source Book for Creative Thinking* (pp. 217–226). New York: Scribner.

Gough, H.G. (1979). A creative personality scale for the Adjective Check List. *Journal of Personality and Social Psychology, 37*:1,398–1,405.

Gough, H.G. (1992). Assessment of creative potential in psychology and development of a creative temperament scale for the CPI. In J.C. Rosen and P. McReynolds, Eds., *Advances in Psychological Assessment* (pp. 225–257). New York: Springer.

Gough, H.G. (1996). *Manual: The California Psychological Inventory* (3rd ed.). Palo Alto, CA: Consulting Psychologists Press.

Gough, H.G., and A.B. Heilbrun, Jr. (1983). *The Adjective Check List Manual*. Palo Alto, CA: Consulting Psychologists Press.

Harrison, M.M., N.L. Neff, A.R. Schwall, and X. Zhao. (2006). *A Meta-analytic Investigation of Individual Creativity and Innovation*. Presented at the Annual Conference for the Society for Industrial and Organizational Psychology, Dallas, TX.

Hezlett, S.A., and N.R. Kuncel. (2012). Prioritizing the learning agility research agenda. *Industrial and Organizational Psychology, 5*(3):296–301.

Hon, A.H.Y., M. Bloom, and J.M. Crant. (2014). Overcoming resistance to change and enhancing creative performance. *Journal of Management, 40*(3):919–941.

Hoover, S.M., and J.F. Feldhusen. (1990). The scientific hypothesis formulation ability of gifted ninth-grade students. *Journal of Educational Psychology, 82*(4):838–848.

Hough, L.M. (1992). The "Big Five" personality variables-construct confusion: Description versus prediction. *Human Performance, 5*(1-2):139–155.

Hough, L.M., and B.S. Connelly. (2012). Personality measurement and use in industrial and organizational psychology. In K.F. Geisinger, Editor-in-Chief, *American Psychological Association Handbook on Testing and Assessment* and N. Kuncel (Vol. 1 ed.), *Test Theory and Testing and Assessment in Industrial and Organizational Psychology* (pp. 501–531). Washington, DC: American Psychological Association.

Hough, L.M., and S. Dilchert. (2007). *Inventors, Innovators, and Their Leaders: Selecting for Conscientiousness Will Keep You "Inside the Box."* Paper presented at the Society for Industrial and Organizational Psychology's 3rd Leading Edge Consortium: Enabling Innovation in Organizations, Kansas City, MO.

Hough, L.M., and J.W. Johnson. (2013). Use and importance of personality variables in work settings. In I.B. Weiner (Ed.-in-Chief) and N. Schmitt and S. Highhouse (Vol. Eds.), *Handbook of Psychology: Vol. 12. Industrial and Organizational Psychology* (pp. 211–243). New York: Wiley & Sons.

Hough, L.M., and D. Ones. (2001). The structure, measurement, validity, and use of personality variables in industrial, work, and organizational psychology. In N.R. Anderson, D.S. Ones, H.K. Sinangil, and C. Viswesvaran, Eds., *Handbook of Industrial, Work, and Organizational Psychology* (pp. 233–277). New York: Sage.

Hough, L.M., F.L. Oswald, and J. Ock. (2015). Beyond the Big Five—A paradigm shift in researching the structure and role of personality. *Annual Review of Organizational Psychology and Organizational Behavior, 2*.

Huang, J.L., A.M. Ryan, K.L. Zabel, and A. Palmer. (2014). Personality and adaptive performance at work: A meta-analytic investigation. *Journal of Applied Psychology, 99*(1):162–179.

Humphreys, L.G. (1986). Commentary. *Journal of Vocational Behavior, 29*(3):421–437.

Ilgen, D.R., J.R. Hollenbeck, M. Johnson, and D. Jundt. (2005). Teams in organizations: From Input-Process-Output models to IMOI models. *Annual Review of Psychology, 56*:517–543.

John, O.P., L.P. Naumann, and C.J. Soto. (2008). Paradigm shift to the integrative Big Five trait taxonomy: History, measurement, and conceptual issues. In O.P. John, R.W. Robins, and L.A. Pervin, Eds., *Handbook of Personality: Theory and Research* (pp. 114–158). New York: Guilford Press.

Judge, T.A., C.A. Higgins, C.J. Thoresen, and M.R. Barrick. (1999). The Big Five personality traits, general mental ability, and career success across the life span. *Personnel Psychology,* 52(3):621–652.

Kell, H.J., D. Lubinski, C.P. Benbow, and J.H. Steiger. (2013). Creativity and technical innovation: Spatial ability's unique role. *Psychological Science,* 24(9):1,831–1,836.

King, N. (1990). Innovation at work: The research literature. In M.A. West and J.L. Farr, Eds., *Innovation and Creativity at Work: Psychology and Organizational Strategies* (pp. 15–59). Oxford, UK: Wiley & Sons.

Koutstaal, W. (2012). *The Agile Mind.* New York: Oxford University Press.

Lombardo, M.M., and R.W. Eichinger. (2000). High potentials as high learners. *Human Resource Management,* 39(4):321–330.

Mason, W., and D.J. Watts. (2012). Collaborative learning in networks. *Proceedings of the National Academy of Sciences of the United States of America,* 109:764–769.

McHenry, J.J., L.M. Hough, J.L. Toquam, M.A. Hanson, and S. Ashworth. (1990). Project A validity results: The relationship between predictor and criterion domains. *Personnel Psychology,* 43(2):335–354.

Milan, L.M., D.M. Bourne, M.M. Zazanis, and P.T. Bartone. (2002). *Measures Collected on the USMA Class of 1998 as Part of the Baseline Officer Longitudinal Data Set (BOLDS)* (ARI Tech. Rep. No. 1127). Alexandria, VA: U.S. Army Research Institute for the Behavioral and Social Sciences.

Mumford, M.D., and S.B. Gustafson. (1988). Creativity syndrome: Integration, application, and innovation. *Psychological Bulletin,* 103(1):27–43.

Mumford, M.D., M.A. Marks, M.S. Connelly, S.J. Zaccaro, and J.F. Johnson. (1998). Domain-based scoring of divergent-thinking tests: Validation evidence in an occupational sample. *Creativity Research Journal,* 11(2):151–163.

Nguyen, N.T., and M.A. McDaniel. (2000). Faking and forced-choice scales in applicant screening: A meta-analysis. Presented at the 15th Annual Meeting of the Society for Industrial and Organizational Psychology, New Orleans, LA.

Oswald, F.L., and L.M. Hough. (2011). Personality and its assessment in organizations: Theoretical and empirical developments. In S. Zedeck, Ed., *APA Handbook of Industrial and Organizational Psychology: Vol. 2. Selecting and Developing Members for the Organization* (pp. 153–184). Washington, DC: American Psychological Association.

Oswald, F.L., and K.L. Schell. (2010). Developing and scaling personality measures: Thurstone was right—but so far, Likert was not wrong. *Industrial and Organizational Psychology,* 3(4):481–484.

Oswald, F.L., L.M. Hough, and J. Ock. (2013). Theoretical and empirical structures of personality: Implications for measurement, modeling and prediction. In N. Christiansen and R. Tett, Eds., *Handbook of Personality at Work* (pp. 11–29). New York: Informa UK Limited/Taylor and Francis Psychology Press.

Pulakos, E.D., S. Arad, M.A. Donovan, and K.E. Plamondon. (2000). Adaptability in the workplace: Development of a taxonomy of adaptive performance. *Journal of Applied Psychology,* 85(4):612–624.

Pulakos, E.D., N. Schmitt, D.W. Dorsey, S. Arad, J.W. Hedge, and W.C. Borman. (2002). Predicting adaptive performance: Further tests of a model of adaptability. *Human Performance,* 15(4):299–323.

Rauch, A., and M. Frese. (2007). Let's put the person back into entrepreneurship research: A meta-analysis on the relationship between business owners' personality traits, business creation, and success. *European Journal of Work and Organizational Psychology,* 16(4):353–385.

Ree, M.J., and J.A. Earles. (1991). Predicting training success: Not much more than g. *Personnel Psychology,* 44(2):321–332.

Rocklin, T. (1994). Relation between typical intellectual engagement and openness: Comment on Goff and Ackerman (1992). *Journal of Educational Psychology,* 86(1):145–149.

Salas, E., D.E. Sims, and C.S Burke. (2005). Is there a "Big Five" in teamwork? *Small Group Research,* 36(5):555–599.

Salgado, J.F., and G. Tauriz. (2014). The Five-Factor Model, forced-choice personality inventories and performance: A comprehensive meta-analysis of academic and occupational studies. *European Journal of Work and Organizational Psychology,* 23(1):3–30.

Saucier, G., and F. Ostendorf. (1999). Hierarchical subcomponents of the Big Five personality factors: A cross-language replication. *Journal of Personality and Social Psychology,* 76:613–627.

Shermis, M.D., and J. Burstein., Eds. (2013). *Handbook of Automated Essay Evaluation: Current Applications and New Directions.* New York: Routledge.

Simonton, D.K. (2012). Taking the U.S. Patent Office criteria seriously: A quantitative three-criterion creativity definition and its implications. *Creativity Research Journal,* 24(2-3):97–106.

Soto, C.J., and O.P. John. (2009). Using the California Psychological Inventory to assess the Big Five personality domains: A hierarchical approach. *Journal of Research in Personality,* 43(1):25–38.

Stark, S. (2002). *A New IRT Approach to Test Construction and Scoring Designed to Reduce the Effects of Faking in Personality Assessment.* (Doctoral dissertation). University of Illinois at Urbana-Champaign. Available: http://psychology.usf.edu/faculty/sestark/ [December 2014].

Stark, S., O.S. Chernyshenko, and F. Drasgow. (2005). An IRT approach to constructing and scoring pairwise preference items involving stimuli on different dimensions: The multi-unidimensional pairwise-preference model. *Applied Psychological Measurement,* 29(3):184–203.

Stark, S., O.S. Chernyshenko, F. Drasgow, C.D. Nye, L.A. White, T. Heffner, and W.L. Farmer. (2014). From ABLE to TAPAS: A new generation of personality tests to support military selection and classification decisions. *Military Psychology,* 26(3):153–164.

Sternberg, R.J. (2001). What is the common thread of creativity? Its dialectical relation to intelligence and wisdom. *American Psychologist,* 56(4):360–362.

Torrance, E.P. (1974). *Norms Technical Manual: Torrance Tests of Creative Thinking.* Lexington, MA: Ginn.

Viswesvaran, C., and D.S. Ones. (1999). Meta-analyses of fakability estimates: Implications for personality measurement. *Educational and Psychological Measurement,* 59(2):197–210.

von Stumm, S., B. Hell, and T. Chamorro-Premuzic. (2011). The hungry mind: Intellectual curiosity is the third pillar of academic performance. *Perspectives on Psychological Science,* 6(6):574–588.

Wallach, M.A., and N. Kogan. (1965). *Modes of Thinking in Young Children: A Study of the Creativity-Intelligence Distinction.* New York: Holt, Rinehart and Winston.

Weiss, D.S. (1981). A multigroup study of personality patterns in creativity. *Perceptual and Motor Skills,* 52(3):735–746.

Welsh, G. (1975). *Creativity and Intelligence: A Personality Approach.* Chapel Hill, NC: Institute for Research in Social Science, University of North Carolina.

West, M.A., and N.R. Anderson. (1996). Innovation in top management teams. *Journal of Applied Psychology,* 81(6):680–693.

Woo, S.E., P.D. Harms, and N.R. Kuncel. (2007). Integrating personality and intelligence: Typical intellectual engagement and need for cognition. *Personality and Individual Differences,* 43(6):1,635–1,639.

Woolley, A.W., C.F. Chabris, A. Pentland, N. Hashmi, and T.W. Malone. (2010). Evidence for a collective intelligence factor in performance of human groups. *Science, 330*(6,004): 686–688.

Zhang, X., and J. Zhou. (2014). Empowering leadership, uncertainty avoidance, trust, and employee creativity: Interaction effects and a mediating mechanism. *Organizational Behavior and Human Decision Processes, 124*(2):150–164.

Zhou, J., and I.J. Hoever. (2014). Research on workplace creativity: A review and redirection. *Annual Review of Organizational Psychology and Organizational Behavior,* 1:333–359.

Section 5

Methods and Methodology

8

Psychometrics and Technology

Committee Conclusion: The military has long been in the forefront of modernized operational adaptive testing. Recent research offers promise for improvements in measurement in a variety of areas, including the application and modeling of forced-choice measurement methods; development of serious gaming; and pursuing Multidimensional Item Response Theory (MIRT), Big Data analytics, and other modern statistical tools for estimating applicant standing on attributes of interest with greater efficiency. Efficiency is a key issue, as the wide range of substantive topics recommended for research in this report may result in proposed additions to the current battery of measures administered for accession purposes. The committee concludes that such advances in measurement and statistical models merit inclusion in a program of basic research with the long-term goal of improving the Army's enlisted accession system.

INTRODUCTION

The U.S. armed forces' historical commitment to develop and improve recruitment, selection, and job classification processes is reflected in a century of research initiatives since World War I to advance psychological measurement methodology and data analytics. The committee encourages the U.S. Army Research Institute for the Behavioral and Social Sciences (ARI) to continue to support this long tradition of research that aims to increase the precision, validity, efficiency, and security of existing assessments; to develop and evaluate new methods for measuring human capabilities; and to explore methods for analyzing the potentially vast amounts of data that these assessments may generate (e.g., machine learning and other analytic methods inspired by the Big Data movement). Although previous chapters of this report describe psychological constructs of interest for recruit selection and assignment, this chapter focuses on research questions related to psychological-assessment measurement methods, emerging assessment technologies, and statistical analysis approaches that are applicable to many types of data. The committee anticipates that the use of modern psychometric and statistical approaches in assessment will yield payoffs in terms of reduced testing time, increased test security, and improved selection and classification, all of which reduce costs and improve human capital in military and organizational settings.

HISTORICAL BACKGROUND

Psychometric research conducted and supported by the U.S. armed forces has influenced measurement and selection practices and yielded a wealth of information about human capabilities. In the domain of cognitive abilities and vocational interests, for example, the Army General Classification Test (Harrell, 1992) predicted performance in military training and deployments during World War II (Flanagan, 1947). In the domain of personality traits, later research by Tupes and Christal (1961) and Digman (1990) included numerous empirical ratings of adjectives in the English lexicon to identify the Big Five taxonomy of personality traits (Goldberg, 1992). These large collective efforts to develop and validate assessments of individual differences sparked advances in the statistical analysis of measures using classical test theory, which in turn provided a foundation for IRT and other modern measurement methods that undergird tests used now for military personnel screening.

The most widely administered and well-known personnel screening test is the Armed Services Vocational Aptitude Battery (ASVAB; Maier, 1993). The original ASVAB was a battery of 10 paper-and-pencil tests of various cognitive abilities, skills, and knowledge that took approximately 4 hours to complete. These tests were constructed, scored, and equated using classical test theory methods, and the scores were combined using simple weighting or regression methods to create composites for selection and classification decisions. For an overview of studies of the factor structure of the ASVAB, see Box 8-1.

In 1992, building on nearly 30 years of basic psychometric researched funded by the Department of Defense and the Office of Naval Research, a new computerized adaptive test (CAT), IRT-based version of the ASVAB was launched, the CAT-ASVAB (see Sands et al., 1999, for an historical review). Although the CAT-ASVAB measured the same constructs, with reliabilities and validities similar to the paper-and-pencil ASVAB, adaptive item selection reduced test length and seat time by nearly 50 percent, allowing for quicker processing of military applicants or, alternatively, the measurement of additional knowledge, skills, abilities, and other attributes (KSAOs). Moreover, the transition to CAT offered benefits in terms of test security and data screening. For example, reducing the number of applicants taking the paper-and-pencil test over time, coupled with the variability in content that adaptive item selection provided, reduced the risk of a sudden and serious test compromise, as did the implementation of methods for thwarting blatant cheating and other attempts to "game the system." Collectively, these features made CAT-ASVAB one of the most psychometrically sound, sophisticated tests ever developed across either military or civilian settings, and it set a high standard for future high-volume personnel screening instruments in the domain of cognitive abilities and skills.

Concurrent with measurement research to prepare for the deployment and maintenance of CAT-ASVAB, ARI conducted a detailed review of military jobs and potential predictors of successful performance under what was called Project A (Campbell, 1990; Campbell and Knapp, 2001). Project A identified eight components of job performance, subsumed under three broad categories now referred to as task performance, citizenship performance, and counterproductive performance (Rotundo and Sackett, 2002). Project A, as well as other military studies (e.g., Motowidlo and Van Scotter, 1994), indicated that although cognitive ability tests such as the ASVAB are among the best predictors of task performance, they only weakly predict citizenship and counterproductive performance, whereas noncognitive variables—those variables that fall under the domains of personality, motivation, and attitudes—tend to exhibit the opposite pattern of predictive relationships.

Recognizing the potential complementary benefits of adding a noncognitive test to CAT-ASVAB for military screening, Army researchers developed and experimented with a personality questionnaire called the Assessment of Background and Life Experiences (ABLE; White et al., 2001), which measured six constructs using the traditional format of single statements, each asking for a response on a four-point Likert scale (Likert, 1932). ABLE scores predicted performance as expected in low-stakes ("research only") settings, but in situations where examinees were motivated to fake "good," substantial score increases and validity decreases were observed (Hough et al., 1990; White and Young, 1998; White et al., 2001). Consequently, researchers began exploring an alternative multidimensional forced-choice (MFC) format for administering items (whereby test takers choose between statements rather than rating a single statement on a scale), along with format and scoring methods that together might address the problem of faking in personality tests, which tends to inflate test scores and reduce validity.

The result was the Assessment of Individual Motivation (AIM) inven-

> **BOX 8-1**
> **Factor Structure of the ASVAB**
>
> The current ASVAB is a test given to all recruits (Powers, 2013), and it measures nine constructs (or factors) fairly well: general science, arithmetic reasoning, word knowledge, paragraph comprehension, mathematics knowledge, electronics information, auto and shop information, mechanical comprehension, and assembling objects. These tests were designed to be essentially unidimensional (Stout, 1987, 1990), so that unidimensional IRT models could be applied for CAT-ASVAB development (see Drasgow and Parsons, 1983). As noted in Chapter 1, the current ASVAB has a number of good measurement characteristics, including the fact that each subscale measures its associated construct well.
>
> The factor structure of the ASVAB was examined by Kass and colleagues (1983). This standardized battery, which tests multiple cognitive abilities, is the primary selection and classification instrument used by all the U.S. military services. The investigators compared their factor structure results with that found for previous ASVAB samples and for previous forms of the ASVAB. In particular, they examined whether the factor structure was similar for racial/ethnic and sex subgroups, to determine the extent of invariance of ASVAB factor structure across these groups. Using data from a sample of more than 98,000 male and female Army applicants, they conducted an exploratory factor analysis and found four factors that accounted for 93 percent of the total variance: verbal ability, speeded performance, quantitative ability, and technical knowledge. In general, the factor analyses for male, female, white, black, and Hispanic subgroups yielded similar

tory (White and Young, 1998), which measures six personality constructs using a MFC format that requires examinees to make "most like me" and "least like me" choices among similarly desirable personality statements that are presented in blocks of four (tetrads). AIM tetrads are scored by assigning 0 to 2 points for each statement; test scores are then based on summing the points across the relevant statements for each construct. According to White and Young (1998), AIM scale scores are only *partially ipsative* (Hicks, 1970), because the number of statements representing each construct varies, respondents are required to endorse only two of four statements in each tetrad, and the nonendorsed statements are assigned intermediate scores. These features introduce enough variation into the AIM total scores to permit normative decision making. Importantly and perhaps as a consequence, the AIM personality measure proved much more resistant to faking than the ABLE personality measure in field research. Unfortunately, the complexity of the format precluded any near-term transition to CAT because there were no psychometric models, such as IRT, that were directly applicable to MFC tetrad responses.

results. The findings provided evidence that the ASVAB's constructs for cognitive abilities were reliable across diverse samples of candidates.

A subsequent reanalysis of the factor structure of the ASVAB compared it with similar aptitude tests (Wothke et al., 1991). In this factor analysis, 46 tests from the Kit of Factor Referenced Cognitive Tests (the Kit) and the 10 ASVAB subtests were administered to a sample of airmen. Because a total of 56 tests were investigated, every examinee did not receive every test. Instead, matrix sampling was used to pair tests. Matrix sampling requires special factor analytic methods (McArdle, 1994). After consideration of descriptive statistics and editing, the data were assembled into a correlation matrix for exploratory and confirmatory factor analysis, which indicated that three factors were required to explain the correlation structure among the ASVAB scores. These three factors were defined as school attainment (for the word knowledge, paragraph comprehension, general science, and mathematics knowledge constructs), speediness (for numerical operations and coding speed, and some of the arithmetic reasoning construct), and technical knowledge (for auto and shop information, mechanical comprehension, and electronics information). The Kit scores required six factors, and the factors used to explain the ASVAB correlations could largely be placed within the factor space of the Kit factors, indicating that the abilities measured by the ASVAB are a subset of the abilities measured by the Kit. These results suggest that future research to enhance selection and classification should focus on abilities not currently measured by the ASVAB, such as those described in Chapter 4 on Spatial Abilities.

NOTE: ASVAB = Armed Services Vocational Aptitude Battery, CAT-ASVAB = computerized adaptive test (IRT-based version of ASVAB), IRT = Item Response Theory.

For the next several years, ARI supported research to increase the validity of the AIM using methods that capitalize on patterns of relationships among item responses and test scores. For example, Drasgow and colleagues (2004) compared the efficacy of predicting attrition among non–high school diploma grad recruits (see Stark et al., 2011; White et al., 2004) based on (a) logistic regression with AIM scale scores, (b) classification-and-regression-tree methods (Breiman et al., 1997), and (c) logistic regression using IRT odds-based scores derived from separately fitting a graded response model (Samejima, 1969) to data for each AIM subscale. There were two noteworthy findings: (1) Computationally intensive classification methodologies could improve the prediction of attrition relative to regression using ordinary scale scores. (2) In accordance with research by Chernyshenko and colleagues (2001), the graded response model generally did not fit the AIM personality data as effectively as the two- and three-parameter logistic IRT models (Birnbaum, 1968) did for cognitive-ability item responses.

This latter finding provided further support for research suggesting that so-called ideal-point models, which were developed and used for attitude measurement (for example, Andrich, 1996; Coombs, 1964; Roberts et al., 1999; 2000), should be similarly considered for personality measurement (Chernyshenko et al., 2001; 2007; Drasgow et al., 2010; Stark et al., 2006). In simple terms, ideal-point models assume that if personality statements are too negative *or* too positive, then respondents will tend to disagree with them. Note that personality statements that are uniquely appropriate to ideal-point models remain a challenge to write and scale (Dalal et al., 2014; Huang and Mead, 2014; Oswald and Schell, 2010). This is partly because most modern test construction and item evaluation practices are imbued with assumptions from traditional IRT models appropriate to cognitive ability (e.g., dominance models, which assume that a person who tends to answer hard items correctly should also be able to answer most easy items correctly). Thus, personality measurement will continue to benefit from research that continuously improves ideal-point IRT models and other modern psychometric tools and measure-development approaches. Consistently superior criterion-related validity over their traditional counterparts is the end goal of such developments.

In 2005, ARI funded a proposal to develop a new personality assessment system that would integrate findings concerning ideal-point modeling, MFC testing, and CAT. The result was the Tailored Adaptive Personality Assessment System (TAPAS; Drasgow et al., 2012), which was developed to measure up to 21 narrow personality factors (dimensions) and military-specific constructs, using a CAT algorithm based on a "multi-unidimensional pairwise-preference" IRT model (Stark, 2002; Stark et al., 2005, 2012a). Respondents are presented with pairs of personality statements, which are similar in their levels of social desirability and extremity, but they are usually different in the constructs they measure; respondents are then asked to select the statement in each pair that is "more like you."

Initial field research with a nonadaptive paper-and-pencil form of this test, known as TAPAS-95s, showed good validities for predicting citizenship and counterproductive performance outcomes with new soldiers (Knapp and Heffner, 2010). Subsequent simulation research investigating various multidimensional pairwise preference CAT designs (see Drasgow et al., 2012; Stark and Chernyshenko, 2007; Stark et al., 2012a, 2012b) affirmed previous findings with the three-parameter logistic model underlying the CAT-ASVAB that adaptive item selection could provide the same accuracy and precision as nonadaptive tests that were nearly twice as long (e.g., Sands et al., 1999). Starting in May 2009, a 13-dimension 108-item TAPAS CAT was administered to Army applicants in military entrance processing stations with a time limit of 30 minutes. This test was later replaced by various 15-dimension 120-item versions with enhanced capabilities to detect rapid, patterned, and random responding in real time, to promote data integrity (Stark et al., 2012c).

Today the U.S. military tests hundreds of thousands of potential recruits annually in 65 military entrance processing stations around the country. The CAT-ASVAB is administered to approximately 400,000 of these applicants, and the scores on four of what are now nine cognitive subtests are used to determine eligibility for enlistment and various assignments. The TAPAS is administered to a subset of the CAT-ASVAB applicants, and scores on a subset of the 15 personality dimensions are used to compute composites of ability and personality based on knowledge ("can do" composite), attitudes ("will do" composite), and experience and willingness to change (adaptability composite; Stark et al., 2014), which are used to make selection decisions and for research on assignment to military occupational specialties (Drasgow et al., 2012; Nye et al., 2012).

The CAT-ASVAB and TAPAS assess ability and personality, respectively, as complementary KSAOs, but they have several psychometric features in common: (1) The IRT CAT algorithms assume that examinees are about "average" at the start of a test, and from then on, essentially tailor subsequent items to examinees' estimated levels on a given ability or trait at a given point to improve measurement precision with fewer items than traditional nonadaptive tests. (2) Examinee trait scores are computed using Bayesian methods that augment the effectiveness of short tests whenever additional examinee data (informative priors) are available. (3) They incorporate technology that hinders and/or flags examinees who appear to be "gaming the test" by quitting early or responding in an inattentive manner.

The CAT-ASVAB and TAPAS also have two primary differences that highlight needs and opportunities for basic research: First, CAT-ASVAB comprises nine cognitive subtests that are individually administered and scored based on the aforementioned unidimensional dominance model, which is a standard model for cognitive ability tests. Correlations among the subtest scores are sizable, as they tend to be between cognitive ability subtests (e.g., $r = .4$ to $.7$). By contrast, the TAPAS measures 13 to 15 personality dimensions based on a multidimensional pairwise preference format, which is based on the ideal-point model discussed previously. Trait scores for TAPAS dimensions are estimated simultaneously using a multidimensional Bayes modal method, and the trait score intercorrelations are significantly lower, as they tend to be for personality (e.g., $r = .10$ to $.45$).

The second difference is that CAT-ASVAB's subtests contain 11 or 16 items each and approximately 2.5 hours total is allowed for completion.[1] TAPAS tests are also adaptive and typically involve 120 or fewer multidi-

---

[1] See http://www.official-asvab.com/whattoexpect_app.htm [December 2014] for additional information on the test format.

mensional pairwise preference items with a 30-minute time limit (Nye et al., 2012).

FUTURE RESEARCH INVESTMENTS

The above discussion illustrates how advances in measurement technology have helped to increase the efficiency and precision of current assessments, which use structured multiple choice and forced-choice formats. Over the next 20 years, advances in computing capabilities will undoubtedly facilitate the development of more sophisticated psychometric models and better methods for combining data from structured assessments with auxiliary information gathered, for example, from personnel records, background questionnaires, social media, and even devices that can capture examinees' physiological data during testing sessions (including, for example, the potential use of biomarkers as described in Chapter 10; biomarkers are discussed in more detail in Appendix C). The proliferation of mobile computing devices and Wi-Fi access will make it possible to test examinees in their natural environments but will also present new challenges for standardization—and thus challenges for the comparability of test scores used for personnel decisions. The emerging field of serious gaming offers potential for measuring examinee KSAOs with less-structured, highly engaging methods, which could yield vast amounts of streaming data that are best analyzed by methods currently used in physics or computer science, rather than methods used in psychology and education. The next sections of this chapter provide a snapshot of developments in psychometric modeling, gaming and simulation, and Big Data analytics, which the committee believes merit serious attention in the Army's long-term research agenda.

Psychometric Modeling

IRT methods provide the mathematical foundation for many of today's most sophisticated structured assessments. IRT models relate the properties of test items (e.g., difficulty/extremity) and examinee trait levels (e.g., KSAOs such as math, verbal, and spatial abilities [see Chapter 4], conscientiousness, emotional stability, and motivation) to the probability of correctly answering or endorsing items. For practical reasons, most large-scale tests have been constructed, scored, and/or evaluated using unidimensional IRT models, which assume that item responding is a function of just one ability or dimension. To obtain a profile of scores representing an examinee's proficiency in several areas, a sequence of unidimensional tests is typically administered, with each being sufficiently long to achieve an acceptable level of reliability. The broader (more heterogeneous) the constructs measured, the more items that are needed.

Research has shown that violating statistical assumptions by applying unidimensional models to tests that unwittingly or intentionally measure weak to moderate secondary dimensions (e.g., measuring mathematical reasoning with word problems that require language proficiency) does not greatly diminish the accuracy of IRT trait scores (Dragsow and Parsons, 1983) or their correlations with outcome variables (e.g., Dragsow, 1982). However, doing so can contribute to biases (also called "differential item and test functioning") that disadvantage subpopulations of examinees who are lower in proficiency on the unaccounted-for secondary dimensions (e.g., Camilli, 1992; Shealy and Stout, 1993). Moreover, when abilities are highly correlated, administering a sequence of unidimensional tests is inefficient; Multidimensional Item Response Theory (MIRT) methods for scoring responses and for selecting items in CATs can reduce the overall number of items administered and increase measurement precision.

MIRT models conceptualize item responding as a function of multiple correlated dimensions (see, for example, Ackerman, 1989; 1991; Reckase, 2009; Reckase and McKinley, 1991; Reckase et al., 1988). Some items are viewed as factorially complex (i.e., they measure more than one dimension), whereas other items are factorially pure (they measure just one dimension). The probability of correct or positive item responses is portrayed as a function of examinee proficiency along multiple dimensions, overall item difficulty/extremity associated with the item content, and the degree to which items are sensitive to variance in proficiency along the dimensions they assess (as indicated by item discrimination coefficients in IRT; e.g., how well they measure, discriminate, or "load on" the intended factors).

MIRT models have value from a purely diagnostic standpoint. They may reveal characteristics that are obscured by unidimensional models and aid item generation and test revision. They may help explain examinee performance (e.g., Cheng, 2009), which is particularly helpful when a test exhibits adverse impact or differential item or test functioning across demographic groups. Their clearest practical benefit for personnel selection, however, lies in potentially improving test efficiency. MIRT scoring methods allow item responses from one ability or trait to serve as *auxiliary* or *collateral information* that informs the responses for other correlated abilities or traits, and this serves to increase overall measurement efficiency and precision (e.g., de la Torre, 2008, 2009; de la Torre and Patz, 2005). MIRT methods therefore not only get more information out of nonadaptive tests, they also reduce the number of items needed in CAT applications. As shown in simulation studies, CATs based on MIRT methods can attain measurement precision goals with even fewer items than unidimensional CATs (Segall, 1996, 2001a; Yao, 2013). Moreover, collateral information provided by data collected before a testing session (e.g., from application blanks or personnel records) can further increase efficiency by providing better starting values for adaptive item selection, and methods that use response *times*, as well as examinee answers, to improve scoring are emerging (e.g., Ranger and Kuhn, 2012; van der Linden, 2008; van der Linden et al., 2010).

One particular class of MIRT models that is growing in popularity due to increased interest in personality and other noncognitive testing is MFC models. One of the earliest was the multi-unidimensional pairwise preference IRT model (Stark, 2002; Stark et al., 2005, 2012b), which de la Torre and colleagues (2012) recently generalized as the PICK and RANK models for preferential choice and rank responses among blocks of statements (e.g., pairs, triplets, tetrads). Another example is the Thurstonian model by Brown and Maydeu-Olivares (2011, 2012, 2013), which can be expressed using an IRT or common factor model parameterization.

These models have been developed for constructing, calibrating, and scoring MFC measures that are intended to reduce response biases, such as socially desirable responding, that are especially prevalent in personnel selection and promotion environments (Hough et al., 1990; Stark et al., 2012a). In these models, examinees must choose or rank statements within each block, based on how well the statements describe, for example, their thoughts, feelings, or actions. However, statements within a block typically represent different constructs, and they are matched on perceived social desirability to make it more difficult for examinees to "fake good" (for example, sometimes examinees have to choose or rank a set of response options where no option is especially desirable). This could be particularly useful in assessments of constructs such as defensive reactivity and emotion regulation, as described in Chapter 6, Hot Cognition.

In contrast to classical test theory scoring methods that historically proved problematic, these model-based MFC methodologies have been shown to yield normative scores that are suitable for inter-individual as well as intra-individual comparisons. This research, however, is still in its early stages. Gaps remain in understanding the intricacies and implications of test construction practices; the capabilities of parameter estimation procedures with tests of different dimensionality, length, and sample size; how to efficiently calibrate item pools, select items, and control exposure of items with CAT; how to create parallel nonadaptive test forms; how to equate alternative test forms; how to test for measurement invariance; and how to judge the seriousness of any specifications or constraints in test construction that are being violated. In short, all of the questions that have been explored for decades with unidimensional IRT models need to be answered for MFC, and more generally, MIRT models. In addition, although the benefits of unidimensional dominance and ideal-point (Coombs, 1964) IRT models for noncognitive testing have been discussed in many papers over the past two decades (e.g., Andrich, 1988, 1996; Drasgow et al., 2010; Roberts et al., 1999; Stark et al., 2006; Tay et al., 2011), there is still much to learn about their use as a basis for MFC applications.

In addition to improving measurement through better models for item responding, test delivery, and scoring, there is a rapidly growing need for methods that can detect aberrant responding, which includes faking or careless responding, and methods for detecting potential item and test compromises that stem from overuse and sudden, outright security breaches. In the 1980s, many heuristic methods for detecting aberrance and item compromise (for reviews, see Hulin et al., 1983; Karabatsos, 2003; Meade and Craig, 2012; Meijer and Sijtsma, 1995) fell into disuse due to the advent of more effective IRT-based methods, which not only flag suspect response patterns but in some cases provide powerful test statistics that can provide benchmarks for simpler methods under different testing conditions (Drasgow, 1982; Drasgow and Levine, 1986; Drasgow et al., 1987, 1991; Levine and Drasgow, 1988). Drasgow and colleagues' (1985) standardized log likelihood statistic (l_z) became one of the more popular early IRT indices because it was effective for detecting spuriously high- and low-ability scores on nonadaptive cognitive tests and because it could be used not only with dichotomous unidimensional IRT models but also with polytomous unidimensional models and multidimensional test batteries. Over time, researchers began exploring noncognitive applications with the goal of detecting faking, untraitedness or random responding, or unspecified person misfit (e.g., Ferrando and Chico, 2001; Reise, 1995; Reise and Flannery, 1996; Zickar and Drasgow, 1996). Researchers also began examining the efficacy of l_z and newer aberrance detection methods with CAT (Egberink et al., 2010; Nering, 1997). By and large, these studies have shown that faking can be difficult to detect because response distortion that is consistent across items is confounded with trait scores. Similarly, because CAT algorithms typically match item extremity to a respondent's trait level, there are too few opportunities to observe inconsistencies between observed and predicted responses to yield adequate power for aberrance detection (Lee et al., 2014). Consequently, as noncognitive tests and CAT applications become more common, new methods for detecting aberrant response patterns will be needed, as will research that examines the tradeoffs of incorporating items into CATs that may reduce test efficiency for the sake of improving detection. (For more information about faking, detection, and potential solutions, see Ziegler et al., 2012.) In the future, it may be possible that potential uses of neuroscience-based measures marking psychological states (as described in Chapter 10) could be one tool for such detection.

A closely related and perhaps even more important research area for the Army is the detection of aberrant responding and test compromise in connection with unproctored Internet testing (Bartram, 2008; International

Test Commission, 2006; Tippins et al., 2006). Although it is highly unlikely that proctored testing at military entrance processing stations and mobile enlistment testing sites will be entirely obviated in the foreseeable future, it may eventually prove advantageous to prescreen applicants or credential existing service members on personal computing devices, just as corporations are accepting unproctored Internet testing as a way of attracting and processing more applicants, credentialing boards are embracing online continuing education, and universities are expanding online course offerings even for advanced degree credits. With mobile computing device capabilities and sales so rapidly increasing, it may simply become a necessity, especially with an all-volunteer workforce, to make pre-enlistment testing as convenient as possible.

The implications are that, in the modern age of testing that includes CAT and unproctored Internet testing, it will be necessary to consider and conceivably adopt some or all of the following approaches:

- Vet individual scores using aberrance detection and verification testing approaches (Segall, 2001b; Tippins et al., 2006; Way, 1998).
- Protect test content by improving item selection and exposure control methods (Barrada et al., 2010; Hsu et al., 2013; Lee et al., 2008).
- Construct and replenish item pools quickly using automatic item generation methods (e.g., Gierl and Lai, 2012; Irvine and Kyllonen, 2002).
- Automatically assemble tests that meet detailed design specifications (e.g., van der Linden, 2005; van der Linden and Diao, 2011; Veldkamp and van der Linden, 2002).
- Monitor item and test properties to detect compromise (Cizek, 1999; McLeod et al., 2003; Segall, 2002; Yi et al., 2008).
- Actively scan Internet blogs, chat rooms, and websites, which provide coaching tips, answer strategies, and realistic or actual items that might point to individual or organized test compromise efforts (Bartram, 2009; Foster, 2009; Guo et al., 2009).

Web-based and mobile CATs—sometimes referred to as eCATs—have been developed to provide efficient, possibly on-demand, screening of examinees in their natural environments and in settings that may not be conducive to traditional forms of test administration (as would be an important consideration for assessments such as situational judgment tests described in the following chapter). Using a Wi-Fi enabled device, examinees can complete a CAT that runs on a remote server or using a mobile application that can be downloaded and run on a tablet computer or smartphone. Such applications are growing rapidly in health care settings because they can be used to assess patients on a variety of physical and psychological well-being indicators just before consultations with health care practitioners, as well as to monitor symptoms and responses to treatments between office visits. (For a prominent example of web-based CAT in health care, readers may consult Cella and colleagues [2010] or Riley and colleagues [2011], who discuss the Patient Reported Outcome Management Information System [PROMIS] initiative funded by the National Institutes of Health.)

Web-based and mobile CATs are also becoming common in workplace contexts. Example uses include screening job seekers for minimal skills before inviting them for an interview, measuring job knowledge or the effects of training, and developing intelligent tutoring systems (Chernyshenko and Stark, in press). Although less common in the military, web-based and mobile CATs have been developed to measure job-related KSAOs among incumbents (e.g., the Computer Adaptive Screening Test, or CAST; see Horgen et al., 2013; Knapp and Pliske, 1986; McBride and Cooper, 1999) and to develop soldier-centered training systems involving, for example, mobile, virtual classrooms and collaborative-scenario training environments (e.g., TRAIN II; Murphy et al., 2013).

In addition to standardization and fairness issues surrounding mobile assessments, which will take many years to explore, an immediate and more obvious concern is the exposure of items that will be used for decision making. However, algorithms are available for CAT item selection that safeguard against high item overuse/overexposure ratios that lead to higher probability of item content breach.

Examining test overlap[2] (Chang and Zhang, 2002; Way, 1998) provides a sense of item exposure. Uniform exposure of items is characteristic of minimal test overlap (Chen et al., 2003). Wang and colleagues (2014) expanded on test overlap, examining the utility of the standard deviation. The authors' analyses conclude that although tests may have similar mean overlap, a smaller standard deviation indicates that the number of shared items between applicants is uniform and that the advantage of retaking the test at a later time is minimized. In addition to optimizing measurement precision, CAT item selection approaches are also considered in terms of the security of the item bank. Barrada and colleagues (2011) found that matching nonstratified item banks with criteria items that had been selected for minimum distance between the respondent's trait level (Li and Schafer, 2005) and the items' difficulty offered greater test security than did use of the Fisher information function (Lord, 1980).

Further investigation will be necessary to determine (a) how CAT item

---

[2] Test overlap = mean of *between-test overlap* (proportion of items on one administration that appear on another) across all possible pairs of respondents.

selection approaches might enhance or compromise test security, (b) what methods can verify and ensure the identity of *e*CAT test takers as well as the security of the testing setting in advance of beginning an *e*CAT, (c) cybersecurity approaches that guard against hacking, and (d) how automatic item generation might enhance security by expanding item pools and continually replacing frequently administered items. A return-on-investment analysis (a comparison of the magnitude and timing of gains from investing in such testing with the magnitude and timing of investment costs) might be a good starting point for considering a web-based *e*CAT, followed by a more complex examination of *e*CAT security, building on customary approaches and new knowledge.

This section has provided a brief review of historical and recent developments in psychometrics that are directly applicable to tests involving structured item formats. The next section delves into technology developments that offer new opportunities for engaging examinees and perhaps reducing the response biases associated with self-report measures. However, the interactive, dynamic nature of these assessments presents challenges in addition to opportunities. To ensure comparability of scores, standardization will need to be addressed, however scores are computed. And Big Data methods will probably be needed to parse the gigabytes of data that each assessment will generate.

Technology

In terms of potential for use in assessment, in contrast with the deep and long-standing tradition of self-report measures and ratings from peers and supervisors, technology advances such as those enabling immersive and realistic simulations and serious gaming provide opportunities for examinees to demonstrate knowledge, emotions, and interactions through their behavior as it is expressed within rich and often realistic scenarios (National Research Council, 2011). This could be especially productive in assessing constructs, such as those described in Sections 2-4 of this report (Chapters 2–7), which may not be effectively or efficiently assessed through standard or even computer adaptive testing. As Landers (2013) described, simulations and serious gaming are related but have some important distinctions.

Simulations, which may involve physical or computer-based re-creation of real-life environments, involve constructed representations of situations in which a task must be reproduced (with potential utility in assessments such as those of spatial abilities, as described in Chapter 4). They typically involve freedom of choice as well as risk and reward. Simulations can involve systems created solely for the purpose of training, or they can use systems that replicate those used in actual practice. One well-known example of the latter is found in the flight simulators that replicate actual instrument panels for various aircraft models. Simulations allow learners to apply their knowledge and to practice important job-related skills in conditions that involve lower risk and possibly lower cost than real-life situations. An important consideration in the design of simulations is the degree to which psychological and physical fidelity to real-life situations can be achieved.

Serious games are similar to simulations but often involve more narrative, and fidelity to real-life situations may be reduced in order to increase user engagement. The U.S. Army currently uses several simulations and serious games as recruitment and training tools (Landers, 2013), but long-term, research would be needed to determine whether gaming experience unduly influences scores and validities of the assessments for high-stakes uses. Through the use of technology in assessments, the collection of vast amounts of data about critical behaviors—those that predict organizational outcomes such as job performance that may in fact closely resemble the outcome (e.g., job performance) itself—will be possible. Furthermore, serious games can confront examinees with unexpected phenomena that require their adaptability (see Chapter 7 for a discussion of individual differences in adaptability) and use feedback in the interest of maintaining or optimizing performance.

Landers (2013) suggested that simulations and serious games offer potential for testing many skills that may be of particular interest to the military: leadership, decision-making, reasoning, spatial ability, persistence, creativity, and particular technical skills (many of which are discussed elsewhere in this report as recommended future research topics). Personality assessment may also be possible using serious games. Just as technology has opened the door to increasingly sophisticated item-administration and precise scoring algorithms (e.g., CAT applications), technology is also changing methods of assessment through powerful advances in simulation and serious gaming. Sydell and colleagues (2013) asked what might be learned from an examination of keystrokes, mouse clicks, repetition of strategies, and response to untimed tasks. As the authors suggested, some of this information may reflect novel predictors of employee outcomes such as performance, satisfaction, and turnover. However, it could introduce contamination associated with environmental influences or irrelevant personal attributes. An integration of simulation and serious gaming with modern psychometric algorithms, based on some combination of IRT and Big Data methods, could be considered as part of a *learning analytics* model that moves past a test of binary correct-versus-incorrect responses, capturing unique and rich sources of information relevant to performance and to the 21st century skills that elude conventional assessment (Bennett, 2010; Redecker and Johannessen, 2013; see also this report's discussions of performance under stress [Chapter 6], adaptive behavior [Chapter 7], and team work behavior [Chapter 5]).

Use of these technologies for assessment is relatively new, but simulation and serious gaming have a longer history as instructional and learning supports. Kevin Corti of PIXEL Learning has been quoted as saying that serious games "will not grow as an industry unless the learning experience is definable, quantifiable, and measurable. Assessment is the future of serious games" (Bente and Brewer, 2009, p. 327). The military is already engaging in such assessments as tests of specific job/task performance in conjunction with training programs, such as performance after military medical training and after flood or fire emergency training on naval ships (Iseli et al., 2010; Koenig et al., 2010; 2013). The reports on these tests provide a framework in terms of scoring systems, performance assessment, and the incorporation of learning from mistakes.

Sydell and colleagues (2013) outlined an assessment approach for simulation and serious gaming that includes identifying what is to be assessed at different levels of the simulated scenario; developing a broad developmental rubric of the measured domain(s), their components, and their relationships with one another (a theoretical model so to speak); application of a Bayesian network (Levy and Mislevy, 2004) that empirically models the probabilistic and dynamic association among measured variables; and developing, vetting, and using various types of score generation tools.

Mislevy advocated the development of simulation-based assessment that is rooted in solid design and psychometrics as opposed to using a data mining approach after the simulation is built (Mislevy, 2013; Mislevy et al., 2012). Mislevy's approach is similar to that of Sydell and colleagues (2013), building a framework referred to as an Evidence-Centered Design, or ECD (Mislevy and Riconscente, 2006; Mislevy et al., 2003). The ECD identifies operational *layers* of the assessment process: domain analysis; domain modeling; specification of a conceptual assessment framework; assessment implementation; and, assessment delivery (see Mislevy, 2013, for a description of the ECD approach). IRT is emphasized as critical to the ECD assessment implementation layer, as are CAT applications in the assessment delivery layer. (For a description of Epistemic Network Analysis, a method for assessing user performance based on ECD, see Shaffer et al., 2009).

New and emerging advances in IRT and constrained optimization procedures will undoubtedly be helpful in developing assessment approaches for simulation and serious gaming. Furthermore, IRT could be used to optimize serious gaming by calibrating scenarios and tasks, using information functions to optimally choose activities for examinees to complete, and using the trait scores to route examinees through gaming levels ranging from novice to expert, much like traditional CAT applications (Batista et al., 2013).

Applications for simulation and serious gaming are growing. Surgical procedures are practiced in simulated venues, as are landing commercial and military aircraft under challenging circumstances, managing business and economic scenarios under unanticipated conditions, and other learning contexts where practice builds skill. Integrated assessment within simulation and serious gaming is gaining traction. Simulation-based assessment focused on patient problems is currently a part of a computer-based medical licensing exam in the United States (Dillon and Clauser, 2009). The Army Research Laboratory uses simulation in its Generalized Intelligent Framework for Tutoring. Still in its infancy is investigating what can be learned about a simulation "player" not only from that player's performance and success in simulation and gaming outcomes but also from the player's keystrokes, mouse clicks, strategy selection (e.g., repetition versus innovation), and time-constrained versus untimed task behavior (see the discussion of performance under stress contained in Chapter 6).

In addition to examining behavioral performance representations, sensors can be used to assess physiological responses and biomarkers, such as galvanic skin response, facial electromyography, electroencephalography, and cardiac activity (Nacke, 2009). (Appendix D describes many of the potentially relevant neuroscience measurement technologies; Chapter 10 has further discussion of potential assessments of individual differences using these technologies.) Examining the use of dynamic Bayesian networks in the contexts of decision making, situational judgment, communication approach, and management of uncertainty has potential value to the Army in making decisions on selection and assignment. In short, although feasibility considerations for large-scale screening would need to be resolved, the committee believes dynamic interactive assessments such as simulation and serious gaming provide two productive investigative areas to better understand what potential recruits actually do in simulated realistic settings as opposed to what they self-report they would do on traditional assessment questionnaires.

RESEARCH RECOMMENDATION

Modern measurement methods come with the promise of increasing precision, validity, efficiency, and security of current, emerging, and future forms of assessment. The U.S. Army Research Institute for the Behavioral and Social Sciences should continue to support developments to advance psychometric methods and data analytics.

A. Potential topics of research on Item Response Theory (IRT) include the use of multidimensional IRT models, the application of rank and preference methods, and the estimation of applicant standing on the attributes of interest with greater efficiency (e.g., via automatic item generation, automated test assembly, detect-

ing item pool compromise, multidimensional test equating, using background information in trait estimation).
B. Ecological momentary assessments (e.g., experience sampling) and dynamic interactive assessments (e.g., team interaction, gaming, and simulation) yield vast amounts of examinee data, and future research should explore the new challenges and opportunities for innovation in psychometric and Big Data analytics.
C. Big Data analytics also may play an increasingly important role as candidate data from multiple diverse sources become increasingly available. Big Data methods designed to find structure in datasets with many more columns (variables) than rows (candidates) might help identify robust variables, important new constructs, interactions between constructs, and nonlinear relationships between those constructs and candidate outcomes.

REFERENCES

Ackerman, T.A. (1989). Unidimensional IRT calibration of compensatory and noncompensatory multidimensional items. *Applied Psychological Measurement, 13*(2):113–127.

Ackerman, T.A. (1991). The use of unidimensional parameter estimates of multidimensional items in adaptive testing. *Applied Psychological Measurement, 15*(1):13–24.

Andrich, D. (1988). The application of an unfolding model of the PIRT type to the measurement of attitude. *Applied Psychological Measurement, 12*(1):33–51.

Andrich, D. (1996). A hyperbolic cosine latent trait model for unfolding polytomous responses: Reconciling Thurstone and Likert methodologies. *British Journal of Mathematical and Statistical Psychology, 49*(2):347–365.

Barrada, J.R., J. Olea, V. Ponsoda, and J. Abad. (2010). A method for comparison of item selection rules in computerized adaptive testing. *Applied Psychological Measurement, 34*(6):438–452.

Barrada, J.R., J. Abad, and J. Olea. (2011). Varying the valuating function and the presentable bank in computerized adaptive testing. *Spanish Journal of Psychology, 14*(1):500–508.

Bartram, D. (2008). The advantages and disadvantages of on-line testing. In S. Cartwright and C.L. Cooper, Eds., *The Oxford Handbook of Personnel Psychology* (pp. 234–260). Oxford, UK: Oxford University Press.

Bartram, D. (2009). The International Test Commission guidelines on computer-based and internet-delivered testing. *Industrial and Organizational Psychology, 2*(1):11–13.

Batista, M.H.E., J.L.V. Barbosa, J.E. Tavares, and J.L. Hackenhaar. (2013). Using the item response theory (IRT) for educational evaluation through games. *International Journal of Information and Communication Technology Education, 9*(3):27–41.

Bennett, R.E. (2010). Technology for large-scale assessment. In P. Peterson, E. Baker, and B. McGaw, Eds., *International Encyclopedia of Education* (3rd ed., vol. 8, pp. 48–55). Oxford, UK: Elsevier.

Bente, G., and J. Breuer. (2009). Making the implicit explicit: Embedding measurement in serious games. In V. Ritterfeld, M. Cody, and P. Vorderer, Eds., *Serious Games: Mechanisms and Effects* (pp. 322–343). New York: Routledge.

Birnbaum, A. (1968). Some latent trait models and their use in inferring an examinee's ability. In F.M. Lord and M.R. Novick, Eds., *Statistical Theories of Mental Test Scores* (pp. 395–479). Reading, PA: Addison-Wesley.

Breiman, L., J. Friedman, R. Olshen, and C. Stone. (1997). *CART* (version 4.0) [Computer program and documentation]. San Diego, CA: Salford Systems.

Brown, A., and A. Maydeu-Olivares. (2011). Item response modeling of forced-choice questionnaires. *Educational and Psychological Measurement, 71*(3):460–502.

Brown, A., and A. Maydeu-Olivares. (2012). Fitting a Thurstonian IRT model to forced-choice data using Mplus. *Behavior Research Methods, 44*(4):1,135–1,147.

Brown, A., and A. Maydeu-Olivares. (2013). How IRT can solve problems of ipsative data in forced-choice questionnaires. *Psychological Methods, 18*(1):36–52.

Camilli, G. (1992). A conceptual analysis of differential item functioning in terms of a multidimensional item response model. *Applied Psychological Measurement, 16*(2):129–147.

Campbell, J.P. (1990). An overview of the Army selection and classification project (Project A). *Personnel Psychology, 43*(2):231–239.

Campbell, J.P., and D.J. Knapp, Eds. (2001). *Exploring the Limits in Personnel Selection and Classification*. Mahwah, NJ: Lawrence Erlbaum Associates.

Cella, D., N. Rothrock, S. Choi, J.S. Lai, S. Yount, and R. Gershon. (2010). PROMIS overview: Development of new tools for measuring health-related quality of life and related outcomes in patients with chronic diseases. *Annals of Behavioral Medicine, 39(Annual Meeting Supplement 1)*:s47.

Chang, H.H., and J. Zhang. (2002). Hypergeometric family and item overlap rates in computerized adaptive testing. *Psychometrika, 67*(3):387–398.

Chen, S.Y., R.D. Ankenmann, and J.A. Spray. (2003). The relationship between item exposure and test overlap in computerized adaptive testing. *Journal of Educational Measurement, 40*(2):129–145.

Cheng, Y. (2009). When cognitive diagnosis meets computerized adaptive testing: CD-CAT. *Psychometrika, 74*(4):619–632.

Chernyshenko, O.S., and S. Stark (in press). Mobile psychological assessment. In F. Drasgow, Ed., *Technology and Testing: Improving Educational and Psychological Measurement, Vol. 2* (NCME Book Series). Hoboken, NJ: Wiley-Blackwell. Available: http://ncme.org/publications/ncme-book-series/ [December 2014].

Chernyshenko, O.S., S. Stark, K.Y. Chan, F. Drasgow, and B.A. Williams. (2001). Fitting Item Response Theory models to two personality inventories: Issues and insights. *Multivariate Behavioral Research, 36*(4):523–562.

Chernyshenko, O.S., S. Stark, F. Drasgow, and B.W. Roberts. (2007). Constructing personality scales under the assumptions of an ideal point response process: Toward increasing the flexibility of personality measures. *Psychological Assessment, 19*(1):88–106.

Cizek, G.J. (1999). *Cheating on Tests: How to Do It, Detect It, and Prevent It*. Mahwah, NJ: Lawrence Erlbaum Associates.

Coombs, C.H. (1964). *A Theory of Data*. New York: Wiley & Sons.

Dalal, D.K., N.T. Carter, and C.J. Lake. (2014). Middle response scale options are inappropriate for ideal point scales. *Journal of Business and Psychology, 29*(3):463–478.

de la Torre, J. (2008). Multidimensional scoring of abilities: The ordered polytomous response case. *Applied Psychological Measurement, 32*(5):355–370.

de la Torre, J. (2009). Improving the quality of ability estimates through multidimensional scoring and incorporation of ancillary variables. *Applied Psychological Measurement, 33*(6):465–485.

de la Torre, J., and R.J. Patz. (2005). Making the most of what we have: A practical application of multidimensional Item Response Theory in test scoring. *Journal of Educational and Behavioral Statistics, 30*(3):295–311.

de la Torre, J., V. Ponsoda, I. Leenen, and P. Hontangas. (2012, April). *Examining the Viability of Recent Models for Forced-Choice Data.* Presented at the Meeting of the American Educational Research Association, Vancouver, British Columbia, Canada. Available: http://www.aera.net/tabid/13128/Default.aspx [February 2015].

Digman, J.M. (1990). Personality structure: Emergence of the five-factor model. *Annual Review of Psychology, 41*:417–440.

Dillon, G.F., and B.E. Clauser. (2009). Computer-delivered patient simulations in the United States Medical Licensing Examination (USMLE). *Simulation in Healthcare, 4*(1):30–34.

Drasgow, F. (1982). Choice of test models for appropriateness measurement. *Applied Psychological Measurement, 6*(3):297–308.

Drasgow, F., and C.K. Parsons. (1983). Application of unidimensional item response theory models to multidimensional data. *Applied Psychological Measurement, 7*(2):189–199.

Drasgow, F., and M.V. Levine. (1986). Optimal detection of certain forms of inappropriate test scores. *Applied Psychological Measurement, 10*(1):59–67.

Drasgow, F., M.V. Levine, and E.A. Williams. (1985). Appropriateness measurement with polychotomous item response models and standardized indices. *British Journal of Mathematical and Statistical Psychology, 38*(1):67–86.

Drasgow, F., M.V. Levine, and M.E. McLaughlin. (1987). Detecting inappropriate test scores with optimal and practical appropriateness indices. *Applied Psychological Measurement, 11*(1):59–79.

Drasgow, F., M.V. Levine, and M.E. McLaughlin. (1991). Appropriateness measurement for multidimensional test batteries. *Applied Psychological Measurement, 15*(2):171–191.

Drasgow, F., W.C. Lee, S. Stark, and O.S. Chernyshenko. (2004). Alternative methodologies for predicting attrition in the Army: The new AIM scales. In D.J. Knapp, E.D. Heggestad, and M.C. Young, Eds., *Understanding and Improving the Assessment of Individual Motivation (AIM) in the Army's GED Plus Program* (pp. 7–1 to 7–16). Arlington, VA: U.S. Army Research Institute for the Behavioral and Social Sciences.

Drasgow, F., O.S. Chernyshenko, and S. Stark. (2010). 75 years after Likert: Thurstone was right (focal article). *Industrial and Organizational Psychology, 3*(4):465–476.

Drasgow, F., S. Stark, O.S. Chernyshenko, C.D. Nye, C.L. Hulin, and L.A. White. (2012). *Development of the Tailored Adaptive Personality Assessment System (TAPAS) to Support Army Selection and Classification Decisions* (Technical Report 1311). Arlington, VA: U.S. Army Research Institute for the Behavioral and Social Sciences.

Egberink, J.L., R.R. Meijer, B.P. Veldkamp, L. Schakel, and N.G. Smid. (2010). Detection of aberrant item score patterns in computerized adaptive testing: An empirical example using the CUSUM. *Personality and Individual Differences, 48*(8):921–925.

Ferrando, P.J., and E. Chico. (2001). Detecting dissimulation in personality test scores: A comparison between person-fit indices and detection scales. *Educational and Psychological Measurement, 61*(6):997–1,012.

Flanagan, J. (1947). Scientific development of the use of human resources: Progress in the Army Air Forces. *Science, 105*(2,716):57–60.

Foster, D. (2009). Secure, online, high-stakes testing: Science fiction or business reality? *Industrial and Organizational Psychology: Perspectives on Science and Practice, 2*(1):31–34.

Gierl, M.J., and H. Lai. (2012). The role of item models in automatic item generation. *International Journal of Testing, 12*(3):273–298.

Goldberg, L.R. (1992). The development of markers of the Big Five factor structure. *Psychological Assessment, 4*(1):26–42.

Guo, J., L. Tay, and F. Drasgow. (2009). Conspiracies and test compromise: An evaluation of the resistance of test systems to small-scale cheating. *International Journal of Testing, 9*(4):283–309.

Harrell, T.W. (1992). Some history of the Army General Classification Test. *Journal of Applied Psychology,* 77(6):875–878.

Hicks, L.E. (1970). Some properties of ipsative, normative, and forced-choice normative measures. *Psychological Bulletin,* 74(3):167–184.

Horgen, K.E., C.D. Nye, L.A. White, K.A. LaPort, R.R. Hoffman, F. Drasgow, O.S. Chernyshenko, S. Stark, and J.S. Conway. (2013). *Validation of the Noncommissioned Officer Special Assignment Battery* (Technical Report 1328). Ft. Belvoir, VA: U.S. Army Research Institute for the Behavioral and Social Sciences.

Hough, L.M., N.K. Eaton, M.D. Dunnette, J.D. Kamp, and R.A. McCloy. (1990). Criterion-related validities of personality constructs and the effect of response distortion on those validities. *Journal of Applied Psychology,* 75(5):581–595.

Hsu, C.L., W.C. Wang, and S.Y. Chen. (2013). Variable length computerized adaptive testing based on cognitive diagnosis models. *Applied Psychological Measurement,* 37(7): 563–582.

Huang, J., and A.D. Mead. (2014). Effect of personality item writing on psychometric properties of ideal-point and Likert scales. *Psychological Assessment.* Epub July 7, available: http://www.ncbi.nlm.nih.gov/pubmed/24999752 [December 2014].

Hulin, C.L., F. Drasgow, and C.K. Parsons. (1983). *Item Response Theory: Application to Psychological Testing.* Homewood, IL: Dow Jones-Irwin.

International Test Commission. (2006). International guidelines on computer-based and Internet delivered testing. *International Journal of Testing,* 6(2):143–172.

Irvine, S.H., and P.C. Kyllonen, Eds. (2002). *Item Generation for Test Development.* Mahwah, NJ: Lawrence Erlbaum Associates.

Iseli, M.R., A.D. Koenig, J.J. Lee, and R. Wainess. (2010). *Automatic Assessment of Complex Task Performance in Games and Simulations* (CRESST Report 775). Los Angeles: University of California, National Center for Research on Evaluation, Standards, and Student Testing.

Karabatsos, G. (2003). Comparing the aberrant response detection performance of thirty-six person-fit statistics. *Applied Measurement in Education,* 16(4):277–298.

Kass, R.A., K.J. Mitchell, F.C. Grafton, and H. Wing. (1983). Factorial Validity of the Armed Services Vocational Aptitude Battery (ASVAB), Forms 8, 9 and 10: 1981 Army Applicant Sample. *Educational and Psychological Measurement,* 43(4):1,077–1,087.

Knapp, D.J., and R.M. Pliske. (1986). *Preliminary Report on a National Cross-Validation of the Computerized Adaptive Screening Test (CAST)* (Research Rep. No. 1430). Alexandria, VA: U.S. Army Research Institute for the Behavioral and Social Sciences.

Knapp, D.J., and T.S. Heffner, Eds. (2010). *Expanded Enlistment Eligibility Metrics (EEEM): Recommendations on a Non-Cognitive Screen for New Soldier Selection* (Technical Report 1267). Arlington, VA: U.S. Army Research Institute for the Behavioral and Social Sciences.

Koenig, A.D., J.J. Lee, M. Iseli, and R. Wainess. (2010). *A Conceptual Framework for Assessing Performance in Games and Simulations* (CRESST Report 771). Los Angeles: University of California, National Center for Research on Evaluation, Standards, and Student Testing.

Koenig, A.D., M. Iseli, R. Wainess, and J.J. Lee. (2013). Assessment methodology for computer-based instructional simulations. *Military Medicine,* 178(10S):47–54.

Landers, R.N. (2013). *Serious Games, Simulations, and Simulation Games: Potential for Use in Candidate Assessment.* Presentation during a data gathering session of the Committee on Measuring Human Capabilities: Performance Potential of Individuals and Collectives, National Research Council. Washington, DC. September 6. Presentation available upon request from the project's public access file.

Lee, Y.H., E.H. Ip, and C-D. Fuh. (2008). A strategy for controlling item exposure in multidimensional computerized adaptive testing. *Educational and Psychological Measurement*, 68(2):215–232.

Lee, P., S. Stark, and O.S. Chernyshenko. (2014). Detecting aberrant responding on unidimensional pairwise preference tests: An application of l_z based on the Zinnes-Griggs ideal point IRT model. *Applied Psychological Measurement*, 38(5):391–403.

Levine, M.V., and F. Drasgow. (1988). Optimal appropriateness measurement. *Psychometrika*, 53(2):161–176.

Levy, R., and R.J. Mislevy. (2004). Specifying and refining a measurement model for a computer-based interactive assessment. *International Journal of Testing*, 4(4):333–369.

Li, Y.H., and W.D. Schafer. (2005). Increasing the homogeneity of CAT's item-exposure rates by minimizing or maximizing varied target functions while assembling shadow tests. *Journal of Educational Measurement*, 42(3):245–269.

Likert, R. (1932). A technique for the measurement of attitudes. In R.S. Woodworth, Ed., *Archives of Psychology* (no. 140, pp. 5–55). New York: Columbia University.

Lord, F.M. (1980). *Applications of Item Response Theory to Practical Testing Problems*. Mahwah, NJ: Lawrence Erlbaum Associates.

Maier, M. (1993). *Military Aptitude Testing: The Past Fifty Years* (DMDC No. 93-007). Monterey, CA: Defense Manpower Data Center.

McArdle, J.J. (1994). Structural factor analysis experiments with incomplete data. *Multivariate Behavioral Research*, 29(4):409–454.

McBride, J.R., and R.R. Cooper. (1999). *Modification of the Computer Adaptive Screening Test (CAST) for Use by Recruiters in All Military Services* (ARI Research Note 99-25). Alexandria, VA: U.S. Army Research Institute for the Behavioral and Social Sciences.

McLeod, L., C. Lewis, and D. Thissen. (2003). A Bayesian method for the detection of item preknowledge in computerized adaptive testing. *Applied Psychological Measurement*, 27(2):121–137.

Meade, A.W., and S.B. Craig. (2012). Identifying careless responses in survey data. *Psychological Methods*, 17(3):437–455.

Meijer, R.R., and K. Sijtsma. (1995). Detection of aberrant item score patterns: A review and new developments. *Applied Measurement in Education*, 8:261–272.

Mislevy, R.J. (2013). Evidence-centered design for simulation-based assessment. *Military Medicine*, 178:107–114.

Mislevy, R.J., and M.M. Riconscente. (2006). *Evidence-Centered Assessment Design: Layers, Structures, and Terminology*. Menlo Park, CA: SRI International.

Mislevy R.J., L.S. Steinberg, and R. Almond. (2003). On the structure of educational assessments. *Measurement: Interdisciplinary Research and Perspectives*, 1(1):3–62.

Mislevy R.J., J.T. Behrens, K.E. Dicerbo, and R. Levy. (2012). Design and discovery in educational assessment: Evidence-centered design, psychometrics, and educational data mining. *Journal of Educational Data Mining*, 4(1):11–48.

Motowidlo, S.J., and J.R. Van Scotter. (1994). Evidence that task performance should be distinguished from contextual performance. *Journal of Applied Psychology*, 79(4):475–480.

Murphy, J., R. Mulvaney, S. Huang, and M.A. Lodato (2013). *Developing Technology-Based Training and Assessment to Support Soldier-Centered Learning*. Presentation at the 28th Annual Conference for the Society of Industrial and Organizational Psychology, Houston, TX.

Nacke, L.E. (2009). *Affective Ludology: Scientific Measurement of User Experience in Interactive Entertainment* (Doctoral dissertation). Blekinge Institute of Technology, Karlskrona, Sweden. Available: http://hci.usask.ca/publications/view.php?id=178 [December 2014].

National Research Council. (2011). *Learning Science Through Computer Games and Simulations*. Committee on Science Learning: Computer Games, Simulations, and Education, M.A. Honey and M.L. Hilton, Eds. Board on Science Education, Division of Behavioral and Social Sciences and Education. Washington, DC: The National Academies Press.

Nering, M.L. (1997). The distribution of indexes of person fit within the computerized adaptive testing environment. *Applied Psychological Measurement*, 21(2):115–127.

Nye, C.D., F. Drasgow, O.S. Chernyshenko, S. Stark, U.C. Kubisiak, L.A. White, and I. Jose. (2012). *Assessing the Tailored Adaptive Personality Assessment System (TAPAS) as an MOS Qualification Instrument* (Technical Report 1312). Ft. Belvoir, VA: U.S. Army Research Institute for the Behavioral and Social Sciences.

Oswald, F.L., and K.S. Schell. (2010). Developing and scaling personality measures: Thurstone was right—but so far, Likert was not wrong. *Industrial and Organizational Psychology*, 3(4):481–484.

Powers, R. (2013). *ASVAB for Dummies: Premier PLUS*. New York: Wiley & Sons.

Ranger, J., and J.T. Kuhn. (2012). Improving Item Response Theory model calibration by considering response times in psychological tests. *Applied Psychological Measurement*, 36(3):214–231.

Rasch, G. (1960). *Probabilistic Models for Some Intelligence and Attainment Tests*. Chicago: University of Chicago Press.

Reckase, M.D. (2009). *Multidimensional Item Response Theory*. New York: Springer-Verlag.

Reckase, M.D., and R.L. McKinley. (1991). The discriminating power of items that measure more than one dimension. *Applied Psychological Measurement*, 15(4):361–373.

Reckase, M.D., T.A. Ackerman, and J.E. Carlson. (1988). Building a unidimensional test using multidimensional items. *Journal of Educational Measurement*, 25(3):193–203.

Redecker, C., and Ø. Johannessen. (2013). Changing assessment—Towards a new assessment paradigm using ICT. *European Journal of Education*, 48(1):79–96.

Reise, S.P. (1995). Scoring method and the detection of person misfit in a personality assessment context. *Applied Psychological Measurement*, 19(3):213–229.

Reise, S.P., and P. Flannery. (1996). Assessing person-fit on measures of typical performance. *Applied Measurement in Education*, 9(1):9–26.

Riley, W.T., P. Pilkonis, and D. Cella. (2011). Application of the National Institutes of Health Patient-reported Outcome Measurement Information System (PROMIS) to mental health research. *The Journal of Mental Health Policy and Economics*, 14(4):201–208.

Roberts, J.S., J.E. Laughlin, and D.H. Wedell. (1999). Validity issues in the Likert and Thurstone approaches to attitude measurement. *Educational and Psychological Measurement*, 59(2):211–233.

Roberts, J.S., J.R. Donoghue, and J.E. Laughlin. (2000). A general item response theory model for unfolding unidimensional polytomous responses. *Applied Psychological Measurement*, 24(1):3–32.

Rotundo, M., and P.R. Sackett. (2002). The relative importance of task, citizenship, and counterproductive performance to global ratings of job performance: A policy-capturing approach. *Journal of Applied Psychology*, 87(1):66–80.

Samejima, F. (1969). Estimation of a latent ability using a response pattern of graded scores. *Psychometrika Monograph Supplement*, No. 17. Available: https://www.psychometricsociety.org/sites/default/files/pdf/MN17.pdf [December 2014].

Sands, W.A., B.K. Waters, and J.R. McBride. (1999). *CATBOOK Computerized Adaptive Testing: From Inquiry to Operation* (No. HUMRRO-FR-EADD-96-26). Alexandria, VA: Human Resources Research Organization.

Segall, D.O. (1996). Multidimensional adaptive testing. *Psychometrika*, 61(2):331–354.

Segall, D.O. (2001a). General ability measurement: An application of multidimensional item response theory. *Psychometrika*, 66(1):79–97.

Segall, D.O. (2001b). *Detecting Test Compromise in High-Stakes Computerized Adaptive Testing: A Verification Testing Approach.* Paper presented at the Annual Meeting of the National Council on Measurement in Education, Seattle, WA. Available: http://ncme.org/default/assets/File/pdf/programPDF/NCMEProgram2001.pdf [February 2015].

Segall, D.O. (2002). An item response model for characterizing test compromise. *Journal of Educational and Behavioral Statistics*, 27(2):163–179.

Shaffer, D.W., D. Hatfield, G.N. Svarovsky, P. Nash, A. Nulty, E. Bagley, K. Frank, A.A. Rupp, and R. Mislevy. (2009). Epistemic network analysis: A prototype for 21st century assessment of learning. *International Journal of Learning and Media*, 1(2):33–53.

Shealy, R., and W. Stout. (1993). A model-based standardization approach that separates true bias/DIF from group ability differences and detects bias/DTF as well as item bias/DIF. *Psychometrika*, 58(2):159–194.

Stark, S. (2002). *A New IRT Approach to Test Construction and Scoring Designed to Reduce the Effects of Faking in Personality Assessment* (Doctoral dissertation). University of Illinois at Urbana-Champaign. Available: http://psychology.usf.edu/faculty/sestark/ [December 2014].

Stark, S., and O.S. Chernyshenko. (2007, October). *Adaptive Testing with the Multi-Unidimensional Pairwise Preference Model.* Paper presented at the 49th Annual Conference of the International Military Testing Association. Gold Coast, Australia. Available: http://www.imta.info/PastConferences/Presentations.aspx?Show=2007 [February 2015].

Stark, S., O.S. Chernyshenko, and F. Drasgow. (2005). An IRT approach to constructing and scoring pairwise preference items involving stimuli on different dimensions: The Multi-Unidimensional Pairwise-Preference Model. *Applied Psychological Measurement*, 29(3):184–203.

Stark, S., O.S. Chernyshenko, F. Drasgow, and B.A. Williams. (2006). Examining assumptions about item responding in personality assessment: Should ideal-point methods be considered for scale development and scoring? *Journal of Applied Psychology*, 91(1):25–39.

Stark, S., O.S. Chernyshenko, W.C. Lee, F. Drasgow, L.A. White, and M.C. Young. (2011). Optimizing prediction of attrition with the U.S. Army's Assessment of Individual Motivation (AIM). *Military Psychology*, 23(2):180–201.

Stark, S., O.S. Chernyshenko, and F. Drasgow. (2012a). Constructing fake-resistant personality tests using item response theory: High stakes personality testing with multidimensional pairwise preferences. In M. Ziegler, C. MacCann, and R.D. Roberts, Eds., *New Perspectives on Faking in Personality Assessments* (pp. 214–239). New York: Oxford University Press.

Stark, S., O.S. Chernyshenko, F. Drasgow, and L.A. White. (2012b). Adaptive testing with multidimensional pairwise preference items: Improving the efficiency of personality and other noncognitive assessments. *Organizational Research Methods*, 15:463–487.

Stark, S., O.S. Chernyshenko, C.D. Nye, F. Drasgow, and L.A. White. (2012c). *Moderators of the Tailored Adaptive Personality Assessment System (TAPAS) Validity.* Ft. Belvoir, VA: U.S. Army Research Institute for the Behavioral and Social Sciences.

Stark, S., O.S. Chernyshenko, F. Drasgow, L.A. White, T. Heffner, C.D. Nye, and W.L. Farmer. (2014). From ABLE to TAPAS: A new generation of personality tests to support military selection and classification decisions. *Military Psychology*, 26(3):153–164.

Stout, W.F. (1987). A nonparametric approach for assessing latent trait unidimensionality. *Psychometrika*, 52:589–617.

Stout, W.F. (1990). A new Item Response Theory modelling approach and applications to unidimensionality assessment and ability estimation. *Psychometrika*, 55:293–325.

Sydell, E., J. Ferrell, J. Carpenter, C. Frost, and C.C. Brodbeck. (2013). Simulation scoring. In M. Fetzer and K. Tyzinski, Eds., *Simulations for Personnel Selection* (pp. 83–107). New York: Springer.

Tay, L., U.S. Ali, F. Drasgow, and B. Williams. (2011). Fitting IRT models to dichotomous and polytomous data: Assessing the relative model–data fit of ideal point and dominance models. *Applied Psychological Measurement*, 35(4):280–295.

Tippins, N.T., J. Beaty, F. Drasgow, W.M. Gibson, K. Pearlman, D.O. Segall, and W. Shepherd. (2006). Unproctored internet testing in employment settings. *Personnel Psychology*, 59(1):189–225.

Tupes, E.C., and R.E. Christal. (1961). *Recurrent Personality Factors Based on Trait Ratings* (Technical Report ASD-TR-61-97). Lackland Air Force Base, TX: Personnel Laboratory, Air Forces Systems Command.

van der Linden, W.J. (2005). Comparison of item-selection methods for adaptive tests with content constraints. *Journal of Educational Measurement*, 42(3):283–302.

van der Linden, W.J. (2008). Using response times for item selection in adaptive testing. *Journal of Educational and Behavioral Statistics*, 33(1):5–20.

van der Linden, W.J., and Q. Diao. (2011). Automated test-form generation. *Journal of Educational Measurement*, 48(2):206–222.

van der Linden, W.J., R.H.K. Entink, and J.P. Fox. (2010). IRT parameter estimation with response times as collateral information. *Applied Psychological Measurement*, 34(5): 327–347.

Veldkamp, B.P., and W.J. van der Linden. (2002). Multidimensional adaptive testing with constraints on test content. *Psychometrika*, 67(4):575–588.

Wang, C., Y. Zheng, and H.H. Chang. (2014). Does standard deviation matter? Using "standard deviation" to quantify security of multistage testing. *Psychometrika*, 79(1):154–174.

Way, W.D. (1998). Protecting the integrity of computerized testing item pools. *Educational Measurement: Issues and Practice*, 17(4):17–27.

White, L.A., and M.C. Young. (1998, August). *Development and Validation of the Assessment of Individual Motivation (AIM)*. Paper presented at the Annual Meeting of the American Psychological Association, San Francisco, CA. Available: http://www.siop.org/tip/back issues/TIPJuly98/burke.aspx [February 2015].

White, L.A., M.C. Young, and M.G. Rumsey. (2001). Assessment of Background and Life Experiences (ABLE) implementation issues and related research. In J.P. Campbell and D.J. Knapp, Eds., *Exploring the Limits in Personnel Selection and Classification* (pp. 526–528). Mahwah, NJ: Lawrence Erlbaum Associates.

White, L.A., M.C. Young, E.D. Heggestad, S. Stark, F Drasgow, and G. Piskator. (2004). *Development of a Non–High School Diploma Graduate Pre-Enlistment Screening Model to Enhance the Future Force*. Arlington, VA: U.S. Army Research Institute for the Behavioral and Social Sciences.

Wothke, W., L.T. Curran, J.W. Augustin, C. Guerrero Jr., R.D. Bock, B.A. Fairbank, and A.H. Gillett. (1991). *Factor Analytic Examination of the Armed Services Vocational Aptitude Battery (ASVAB) and the Kit of Factor-Referenced Tests (AFHRL-TL-90-67)*. Brooks Air Force Base, TX: Air Force Human Resources Laboratory.

Yao, L. (2013). Comparing the performance of five multidimensional CAT selection procedures with different stopping rules. *Applied Psychological Measurement*, 37(1):3–23.

Yi, Q., J. Zhang, and H.H. Chang. (2008). Severity of organized item theft in computerized adaptive testing: A simulation study. *Applied Psychological Measurement*, 32(3):543–558.

Zickar, M.J., and F. Drasgow. (1996). Detecting faking on a personality instrument using appropriateness measurement. *Applied Psychological Measurement*, 20(1):71–87.

Ziegler, M., C. MacCann, and R.D. Roberts, Eds. (2012). *New Perspectives on Faking in Personality Assessments*. New York: Oxford University Press.

9

Situations and Situational Judgment Tests

Committee Conclusion: The ability to use judgment to interpret, evaluate, and weigh alternate courses of action appropriately and effectively is relevant to a wide variety of situations within the military. Various streams of research, including new conceptual and measurement developments in assessing situational judgment, as well as evidence of consistent incremental validity of situational judgment measures over cognitive ability and personality measures for predicting performance in various work settings, lead the committee to conclude that measures of situational judgment merit inclusion in a program of basic research with the long-term goal of improving the Army's enlisted accession system.

Situational judgment tests (SJTs) are psychological measures that present test takers with hypothetical situations that often reflect constructs that may be interpersonal (e.g., communication, teamwork), intrapersonal (e.g., emotional stability, adaptability), or intellectual (e.g., technical knowledge, continuous learning) in nature. A sample SJT question dealing with squad leadership follows (Hanson and Borman, 1995; for another military SJT measure, see Tucker et al., 2010):

> You are a squad leader on a field exercise, and your squad is ready to bed down for the night. The tent has not been put up yet, and nobody in the squad wants to put up the tent. They all know that it would be the best place to sleep since it may rain, but they are tired and just want to go to bed. What should you do?

A. Tell them that the first four men to volunteer to put up the tent will get light duty tomorrow.
B. Make the squad sleep without tents.
C. Tell them that they will all work together and put up the tent.
D. Explain that you are sympathetic with their fatigue, but the tent must be put up before they bed down.

There are multiple ways to answer an SJT (e.g., pick the best/worst; respond to each option on a 1-5 scale of effectiveness), and furthermore, there is more than one way to score an SJT (two scoring options, among others, are agreement with subject matter expert responses and agreement with the consensus response). SJTs are historically and substantively related to tests of practical intelligence, tacit knowledge, and other tests that ask respondents about solving hypothetical problems one might face in the real world. In fact, the specific test items for measures of practical intelligence look a lot like SJT items (e.g., the Wagner and Sternberg, 1991, measure of the practical intelligence of managers).[1] (For a useful overview of critical SJT characteristics, such as constructs assessed, situational content, scoring methods, instructions and testing medium, see Campion et al., 2014.)

VALIDITY

Criterion-Related Validity

Since its development for use in personnel selection decades ago (e.g., Motowidlo et al., 1990), testing situational judgment has remained a viable method for assessing psychological constructs due to its consistent, criterion-related validity across a variety of work settings. But an SJT is a method of measurement, not a construct (see Arthur and Villado, 2008), and in fact, a variety of constructs have been (and can be) measured with an SJT. Existing SJTs predict task performance and other outcomes where cognitive ability is required (McDaniel et al., 2001); they also have shown validity for organizational citizenship behavior and personality-relevant outcomes where cognitive ability is not a strong requirement (Christian et al., 2010).

SJT items have successfully measured the constructs of job knowledge, interpersonal and team skills, leadership, and personality; meta-analyses of these types of SJTs have found high validities for a variety of types of performance that are technical and interpersonal in nature. Building on early SJT research that focused on constructs such as social and practical intel-

---

[1] Of note, practical intelligence measures tend to not be useful if they are redundant with cognitive measures (Rumsey and Arabian, 2014).

ligence (see Whetzel and McDaniel, 2009), more modern SJTs frequently target interpersonal skills and competencies that are presumably difficult to measure with traditional tests of personality or attitudes. Examples of interpersonal constructs measured by SJTs include organization, maturity, and respectfulness (Weekly and Jones, 1997), work commitment and work quality (Chan and Schmitt, 1997), conflict management and resolution (Chan and Schmitt, 1997; Olson-Buchanan et al., 1998; Richman-Hirsch et al., 2000), leadership (Bergman et al., 2006), communication skills (Lievens and Sackett, 2006), integrity (Becker, 2005), and team orientation (Mumford et al., 2008; Weekley and Jones, 1997).

Discriminant Validity

Even though SJTs have a history of being developed around specific constructs intended to predict specific types of job performance, sometimes they predict unintended outcomes as well or better than the intended outcomes, as was found in a recent meta-analysis (Christian et al., 2010). This meta-analysis found an interpersonal skills SJT predicted the technical aspects of job performance just as well as the interpersonal (contextual) aspects of job performance (meta-analytic correlations of $r = .25$ and $.21$, respectively). This finding of equivalent levels of prediction for unintended outcomes might reflect a need to refine the outcome measures, as much or more than refining the SJTs themselves.

There is a potentially more promising approach to designing SJTs such that they predict intended outcomes more strongly than predicting unintended outcomes. This approach involves designing an SJT so that the possible responses to a given situation each reflect different constructs. Under this design all of the SJT responses, across all situations, can be analyzed using a multitrait-multimethod framework (where traits = SJT responses, and methods = SJT situations/stems in which the responses are nested). A recent SJT designed in this manner successfully partitioned the variance of SJT items between situations and three constructs that measured the tendencies to approach new goals, to avoid new goals, or to treat new goals as achievements that others will evaluate (Westring et al., 2009). This innovative format could help an SJT approach to measure other constructs better, both conceptually and psychometrically. It also improves understanding of the nature of the situation being tested by a particular SJT.

In a related innovative approach, existing Multidimensional Item Response Theory models can be applied to situational judgment measures that are designed in a similar manner. A concrete example, similar to the previous example but designed to test goal orientation, would be an SJT that asks examinees how they would respond to 12 performance situations, where each situation is followed by a set of four possible responses (items), with each response in a set reflecting one of four different constructs: (a) work commitment, (b) work quality, (c) conflict management, and (d) empathy (Chan and Schmitt, 1997). Examinees would pick the best and worst response or rank the responses from most to least effective. This SJT format is highly compatible with Item Response Theory (IRT) methods for investigating patterns of correlations between the four constructs (e.g., Brown and Maydeu-Olivares, 2011, 2013; de la Torre et al., 2012).

Incremental Validity

Testing situational judgment has maintained its decades-long prominence in the research and practice of employment testing because, even though the test formats and the constructs they assess vary widely, SJTs have demonstrated persistent incremental validity above cognitive ability and personality measures when predicting job performance of either a technical or interpersonal nature. Thus, even though particular SJTs are known to be correlated with traditional measures of cognitive ability and personality, as previously discussed, SJTs assess compound knowledge, skills, and abilities that often do not fall clearly or cleanly in either category (e.g., time management skills, leadership behaviors). SJTs are shown to predict above and beyond measures of cognitive ability and personality when SJTs are administered in job applicant settings (Chan and Schmitt, 2002; Clevenger et al., 2001) and in academic settings relevant to college admissions (Oswald et al., 2004). These increments are often modest (Peterson et al., 1999), and they critically depend on the type of SJT, the nature of the criteria being predicted, and the other types of predictors being administered. Nonetheless, the findings of these three studies illustrate how carefully developed SJTs, when added to an existing test battery, can improve selection decisions in the aggregate, across large applicant pools and/or over time. These consistent findings in the literature suggest the potential for increased validity with additional investments in SJT research. Furthermore, the committee predicts that when SJTs and personality assessments "compete" for validity, the military or other organizations might decide in favor of administering SJTs over personality tests (due to, for example, their greater face validity).

INSTRUCTIONS AND FORMAT

"Would Do" versus "Should Do" Instructions

Meta-analytic research indicates that the type of instructions used by an SJT partially determines its relationships with measures of cognitive ability or personality (McDaniel et al., 2007). For instance, an SJT that requires respondents to indicate what one "should do" or to rank-order situational responses by their effectiveness is more highly correlated with cognitive ability than with personality ($r = .32$ versus $r = .10–.20$; also see Lievens et al., 2009). Conversely, the responses to SJTs asking about the respondent's behavioral tendencies or what one "would do" in a situation tend to be more correlated with the personality traits of agreeableness, emotional stability, and conscientiousness (also see the McDaniel and Nguyen, 2001, meta-analysis) and less correlated with cognitive ability ($r = .30–.33$ for the aforementioned constructs versus $r = .17$ for cognitive ability).

That said, there are at least two important qualifications to these general findings. First, whether one provides "would do" versus "should do" instructions for an SJT is somewhat governed by the constructs being measured (e.g., SJTs related to personality tend to ask about behavioral tendencies or what one "would do"), although research indicates that in those cases where the construct and item content were held constant across different instruction sets, validity patterns appeared to be similar (McDaniel et al., 2007). Second, an additional concern might be that in high-stakes operational settings, "would do" instruction sets might lead to inflated mean SJT scores compared with "should do" instruction sets; however, at least one large-sample study in a college admissions setting (Lievens et al., 2009) found no such mean differences for its SJT measuring interpersonal skills.

Video versus Written SJTs

Traditionally, SJTs have been administered in a written format, although there are some notable instances of using video and computerized formats of the SJT, in which the nontraditional format has served at least four important purposes. First, the reading level required for video SJTs is often lower, meaning that if verbal ability is irrelevant to the constructs of interest, then the video SJT can lead to more reliable measurement than its text-based counterpart. This appears to be true not only for samples of test takers that vary widely in verbal ability (Chan and Schmitt, 1997) but also in samples presumed to have higher levels of verbal ability (e.g., medical school applicants; Lievens and Sackett, 2006). Lower verbal-ability requirements also tend to mean lower potential for adverse minority impact (large subgroup differences).

Second, in addition to reducing the demands on verbal ability, video formats are more immersive and engaging testing experiences, hence the enthusiasm for multimedia testing since the ready availability of personal computers in the 1990s. Video SJTs have the potential to allow organizations to distinguish themselves from competitors, to send a signal about their innovation, to pique the interest of highly valued applicants, and to gather information about these applicants for making hiring decisions. Note that video SJTs have equal potential for creating negative impressions about an organization if they are not carefully constructed and administered.

Third, by customizing video SJTs to particular types of work, organizations can provide standardized realistic job previews (Weekley and Jones, 1997) that allow applicants to draw conclusions about person-environment fit (e.g., Edwards and Cable, 2009), which has implications for greater job satisfaction, lower turnover, and longer-term commitment. To the extent these conclusions are accurate, this benefits both applicants and organizations alike.

Fourth, video SJTs offer hope for increasing job applicants' perceptions of the fairness and validity of selection systems. Watching enactments of workplace situations, rather than reading about them, might increase examinee motivation and decrease general cognitive ability requirements, which in turn might serve to reduce the risk of adverse impact on minority candidates. Olson-Buchanan and colleagues (1998) developed video SJTs that used a branching algorithm to present different scenes depending on examinee responses. This interactivity reportedly increased realism because the scenes that ensued were logical consequences of the examinees' choices. Although branching was not performed using an IRT adaptive testing algorithm, as was the case with the researchers' verbal skills assessment, the logic was in fact quite similar to that of the IRT algorithm.

Regarding the nature of video SJTs—in general and when compared with their written counterparts—the body of evidence accumulated over the past decade is complex. Contrary to the findings of Chan and Schmitt (1997) and Richman-Hirsch and colleagues (2000), Lievens and Sackett (2006) found no statistically significant difference in face validity perceptions for video and written SJTs in their study with medical school applicants. Likewise, although Chan and Schmitt (1997) and Olson-Buchanan and colleagues (1998) found smaller ethnic-group differences with video SJTs than written SJTs, Weekley and Jones (1997) found that video SJTs still exhibited considerable subgroup mean differences (0.3 to 0.6 standard deviations) in two studies of hourly service workers.

Like the SJT itself, the video format is a vehicle for measuring a variety of psychological constructs, some of which might be more amenable to the format (e.g., teamwork) than others (e.g., computer programming knowledge). Coupled with potential benefits are potential challenges that come with large-scale administration of video SJTs; these latter challenges may well be mitigated by future testing and video technologies.

Single-Item SJTs

The single-item SJT is a novel and simpler SJT format that has recently emerged in the research literature (Krumm et al., 2014). It amounts to asking test takers to rate for effectiveness a representative set of critical incidents derived from a job analysis. Below are examples of two critical incidents generated by human factors professionals (HFPs) that have been rated on a Likert scale for effectiveness (Motowidlo et al., 2013, p. 1,854):

> Whenever an HFP would perform product testing with live participants, the HFP would invite the entire product management team to the lab to observe the testing. [This example is intended to represent an effective incident]

> The team was discussing how to design an interface. With the exception of the HFP, the team was unanimous in their design. The HFP began raising his voice and telling the team about his educational credentials. After the decision was made to use the design the rest of the team developed, the HFP aggressively stormed out of the room. [This example is intended to represent an ineffective incident]

Single-item SJTs of this nature yield somewhat distinct factors for knowledge of effective versus ineffective situations, and both factors demonstrate validity for predicting performance-related outcomes across samples of job incumbents and undergraduates (Crook et al., 2011; Motowidlo et al., 2013). The validity might be surprising, given the seeming obviousness of the effectiveness or ineffectiveness of situations like the above examples. But validity might emerge because of this obviousness: when examinees cannot identify effective and ineffective critical incidents, this is predictive of important outcomes. In any case, all of these recent empirical findings indicate that there is potential for validity of single-item SJTs in job applicant samples.

In addition, it seems likely that a large number of single-item SJTs could be developed and refined in the same amount of time it typically takes to develop and refine a set of, say, 50 traditional SJT items. Having a large number of SJT items, in turn, increases the potential to generate different SJT forms with similar psychometric qualities. This benefit seems essential in large-scale applications where test forms need to be continuously refreshed to minimize blatant forms of cheating. Still, additional research on single-item SJTs is needed to assess whether validities can be preserved under conditions where job applicants are coached on the correct responses and/or have the motivation to fake.

SUBGROUP DIFFERENCES AND ADVERSE IMPACT

A meta-analysis of subgroup differences by Whetzel and colleagues (2008) found that SJTs collectively exhibit smaller white-black effect size differences than traditional cognitive ability tests (standard deviation of

0.38 versus 1.0), but the magnitudes of the differences depend on cognitive load, which some research has linked to the instructions accompanying SJT scenarios. More specifically, "should do" or knowledge instructions, which ask respondents to choose the best/worst or most/least effective option(s) from a series of alternatives, tend to increase correlations with cognitive ability measures and thus increase subgroup differences. In contrast, "would do" or behavioral tendency instructions, which ask respondents to choose the most/least likely option(s) from a series of alternatives, have reduced subgroup differences relative to "should do" instructions.

However, there is a potential tradeoff with "would do" instructions because even if subgroup differences are reduced, "would do" instructions have also been found to be more susceptible to faking (e.g., McDaniel et al., 2007; Nguyen et al., 2005; Peeters and Lievens, 2005; Ployhart and Ehrhart, 2003). Generally speaking, mean differences between subgroups by race and gender on SJTs tend to be low, but nonetheless, the cited research literature suggests that the magnitude of these differences (and thus the contribution of an SJT to adverse impact in a selection battery) is influenced by both the constructs and the instructions associated with a given SJT. Future research might also investigate subgroup differences in the SJT response process. Because this process involves comprehension, retrieval, judgment, and response selection (Ployhart, 2006), it must be affected by cognitive influences (e.g., verbal complexity) and noncognitive effects (e.g., test-taking anxiety, test-taking motivation) on which subgroups are already known to differ.

PSYCHOMETRIC FINDINGS

Reliability

Typically, psychological measures are developed to be unidimensional or "construct-pure," meaning that the relevant content for all the items within a given scale should reflect a single construct. Alpha reliability coefficients also depend on this assumption to be informative (Cortina, 1993). Developing SJT measures is challenging from the perspectives of both test development and psychometrics because SJT items are a priori known to be complex and heterogeneous in terms of both the stem (situation) of each item and the items' response options. The previous research notwithstanding, SJTs in practice are often quite heterogeneous and will often yield a single weak factor (e.g., McDaniel and Whetzel, 2005; Whetzel and McDaniel, 2009) that is sometimes tautologically labeled "situational judgment." Directly as a function of finding a weak factor, coefficient alpha is notoriously low (Chan and Schmitt, 1997; Smiderle et al., 1994; Weekley and Jones, 1997). By contrast, test-retest reliabilities based on the same items control for item heterogeneity, practice effects notwithstanding. Test-retest reliabilities are potentially higher, suggesting there is reliable variance in SJT items that is not part of an overall construct but instead tends to be unique to each item. For example, the 56 alpha coefficients located by Catano and colleagues (2012) yielded an average value of $\alpha = .46$. In their two longitudinal SJT studies, similarly low alphas were obtained. However, their test-retest reliabilities were much higher ($r = .82$) in a student sample after a 2-week retest interval and in an HR sample ($r = .66$) after a 3-month retest interval (Catano et al., 2012).

Thus, alpha reliability coefficients are inappropriate indices of reliability for SJTs because the situational stems and item responses reflect complex situations and thus are not internally consistent. Test-retest reliability coefficients are more appropriate to show stability in situational judgment across items, but equally important, if not more so, is the value and necessity in tying the content of SJTs to psychological theory and to the information provided by a job analysis. This, together with psychometric evidence for test-retest reliability, provides the converging pieces of evidence that help establish the quality and validity of an SJT measure. In today's age of Big Data analytics and flexible predictive models, a temptation to be avoided in SJT development would be to select items that simply predict based on statistical properties, without theoretical basis or concern for reliability or construct validity. McDonald (1999, p. 243) also refers to this approach disparagingly, noting that it simply "uses the relations of the unique parts of the items to the criterion to maximize predictive ability" so that predictors are chosen that correlate highly with the criterion but correlate near zero with each other. In other words, using validity as a driver for item selection might lead to high levels of prediction by design, but the unfortunate result might be not knowing what is being measured.

Scoring

Not surprisingly, scoring methods have been shown to influence the validity of SJTs for predicting various workplace criteria (Bergman et al., 2006; Weekley and Jones, 1997). As with personality measures, rational scoring keys determine the "correct" answers by theory, and therefore rational keys are the most straightforward to develop. However, rational scoring may increase susceptibility to faking because the "correct" answers may be too obvious.

The alternative approach of empirical keying, where SJT item responses that correlate highest (in magnitude and sign) with a criterion are deemed to be the correct answer, poses different challenges. Depending on the criterion and the group that is used for development, one particular SJT can have many empirical keys, which may include different numbers of items and/or differentially weighted items. For example, empirical keys developed using the consensus judgments of novices (e.g., new employees), experts (e.g., supervisors, managers), customers, or examinees may be substantially different (Bergman et al., 2006). Furthermore, items that differentiate well among novices may be nondiscriminating among experts, and items may be more or less discriminating within a group, depending on the attribute that is being keyed. Because SJT scenarios and response options are typically multidimensional, the actions or events can be evaluated along multiple dimensions. With video SJTs, the inextricable nonverbal cues may lead to differences from written tests that were designed to be the content-equivalent of the video version (see, for example, Weekly and Jones, 1997). Ultimately, the potentially large number of empirical keys, the variations in item properties across calibration samples, and the loss of information when items are discarded due to low discrimination have implications for reliability, validity, and the generalizability of findings.

Parallel Forms

There have been recent attempts to create SJT forms that are reasonably parallel in their overall content and psychometric properties. As a step toward parallel SJT forms construction, researchers have explored some intuitive approaches involving assignment of items to different SJT forms (Irvine and Kyllonen, 2002; Lievens and Sackett, 2007; Oswald et al., 2005; Whetzel and McDaniel, 2009). For example, under an *incident isomorphism* strategy, a large pool of critical incidents is generated; two items, for example, are written for each incident; and one item from each pair is assigned to each form. With *random assignment* strategy, a large pool of items is developed for each domain and the items are assigned randomly to different forms.

Extending the latter approach, Oswald and colleagues (2005) used *stratified random assignment* by assigning SJT items randomly within each of 12 dimensions within intellectual, interpersonal, and intrapersonal domains. This assignment method, in conjunction with traditional scale construction and evaluation practices, produced 144 forms of an SJT to predict different types of college student performance. More specifically, the SJT item means, standard deviations, item-total correlations, and item validities (correlations of item responses with first-year grade point average) were arrayed in a spreadsheet (items in rows, statistics in columns) and grouped by the 12 dimensions they represented. Next, a computer program generated 10,000 preliminary 36-item forms via stratified random sampling of 3 items from each of the 12 dimensions. Projected test form means, standard deviations, and criterion-related validities were computed using widely available formulas. Then, the number of forms was reduced by imposing statistical constraints: The individual test form means could differ by no more than 0.05 standard deviations from the overall mean across all test forms; test form alpha reliabilities had to exceed .70; and test form criterion-related validities had to exceed .15. This process left 144 forms that exhibited only about 30 percent overlapping content. Note that sample sizes were large enough to suggest the item statistics were stable (i.e., $N = 644$ and 381 across two sets of items) and that the 144 forms selected were not unduly capitalizing on chance. But that said, future investigation of the stability of parallel form generation procedures for SJTs is generally recommended.

At the very least, the aforementioned effort illustrates several important points or principles to be examined further in future research: (1) SJTs can be built to exhibit adequate reliabilities and validities by paying attention to statistical indices of homogeneity and criterion validity during test assembly, even while sampling items from conceptually heterogeneous dimensions in a deliberate manner. (2) Alternative SJT forms can be produced in large numbers by using computationally intensive methods that are similar in spirit to adaptive testing and automated test assembly algorithms (e.g., van der Linden, 1998; van der Linden and Glas, 2010) that select items to satisfy multiple constraints such as those tied to content, information, and exposure constraints.

Additionally, new SJT formats might be more amenable to suitably fitting IRT models, such that parallel forms could be constructed by matching test response and information functions (Hulin et al., 1983), test forms could be equated using traditional linking methods (Kolen and Brennan, 2004), measurement invariance tests might be conducted across examinee subpopulations (Millsap and Yun-Tein, 2004), and adaptive testing could be used to improve measurement precision while holding SJT test length constant (Hulin et al., 1983; van der Linden and Glas, 2010).

FUTURE DIRECTIONS

Operational Use of Innovative SJTs

For testing of situational judgment to be useful operationally in personnel selection settings (usually the intended setting), there is a set of key desiderata that are typical for most selection tests (see, for example, the nine research recommendations of Whetzel and McDaniel, 2009): (a) appropriate scoring methods (e.g., empirical versus rational); (b) high reliability, validity, and incremental validity when considering supplementing or substituting measures in a selection battery; (c) low subgroup differences (and adverse impact) with respect to legally protected subgroups; (d) resistance to faking and coaching; (e) ability to create parallel forms; (f) consideration of the possibilities and implications of applicant retesting;

and (g) item and test security. To this end, a strong basic research agenda would be required to (1) examine the relationships of SJT scores with key individual-differences variables; (2) clarify the complex associations between SJT testing modalities, instruction sets, constructs, criterion-related validity, and subgroup differences; and (3) explore new technologies for capturing and scoring examinee responses.

Multimedia and SJTs

Regarding this latter point concerning technology and SJTs, the committee believes there is still a great deal to be learned about the benefits of video and, more generally, multimedia SJTs with respect to lower-fidelity, written alternatives. Technological advances will surely create new possibilities for SJT item and test development, test delivery, response capture, and scoring. For instance, technology has evolved to the extent that multimedia tests can now be administered on a variety of personal computing devices, including tablets and smartphones. This enables testing to take place in natural environments (e.g., at home and unproctored) as well as traditional ones (e.g., at a testing center and proctored). Furthermore, rather than limiting responses to simple mouse clicks and key strokes to answer Likert scales or check-boxes, technology can be employed to seek to collect and analyze open-ended responses, such as by using dictation software or a video camera, with verbal responses analyzed with natural language processing technology (e.g., Jurafsky and Martin, 2008; Kumar, 2011). Accordingly, as new opportunities for testing situational judgment develop, future research will need to examine the advantages and disadvantages of new methodologies from both examinee and institutional perspectives.

Advances in SJT Item Development and Scoring

Given current SJT construction and scoring practices and the emergent nature of IRT methodologies, the committee believes the military testing process would benefit from future research and thinking on how psychometric technology, which has been used to improve precision and efficiency in large-scale testing programs, might be adapted for testing situational judgment. Additionally, the committee encourages applied researchers to think creatively about SJT item writing and answer formats; if changes can be made to increase the suitability of prevailing psychometric models without destroying the realism of SJT items, then an array of new possibilities for test construction and evaluation can be expected to follow.

Closing Point

Organizational researchers are clearly concerned about understanding the nature of testing for situational judgment, both in terms of constructs measured and methods employed in various SJTs. A program of basic SJT research committed to integrating these concerns would likely increase the effectiveness of personnel selection and classification systems.

RESEARCH RECOMMENDATION

The U.S. Army Research Institute for the Behavioral and Social Sciences should support research to understand constructs and assessment methods specific to the domain of situational judgment, including but not limited to the following lines of inquiry:

A. Develop situational judgment tests with items reflecting constructs that are otherwise difficult to assess using other tests, that are important, and that show promise for validity (e.g., prosocial knowledge, team effectiveness).
B. Consider innovative formats for presenting situations (e.g., ranging from simple text-based scenarios to dynamic and immersive computer-generated graphics), capturing examinee responses (e.g., open-ended, voice, gestures, facial expressions, eye movements, reaction times), and evaluating examinee responses (e.g., advanced natural language processing, automated reasoning, machine learning).
C. Develop and explore psychometric models and methods that can accommodate the rich array of data that innovative assessment methods for situational judgment may yield, facilitating the development of psychometrically and practically equivalent assessments, and improving reliability and testing efficiency.

REFERENCES

Arthur, W., Jr., and A.J. Villado. (2008). The importance of distinguishing between constructs and methods when comparing predictors in personnel selection research and practice. *Journal of Applied Psychology, 93*(2):435–442.

Becker, T.E. (2005). Development and validation of a situational judgment test of employee integrity. *International Journal of Selection and Assessment, 13*(3):225–232.

Bergman, M.E., F. Drasgow, M.A. Donovan, J.B. Henning, and S. Juraska. (2006). Scoring situational judgment tests: Once you get the data, your troubles begin. *International Journal of Selection and Assessment, 14*(3):223–235.

Brown, A., and A. Maydeu-Olivares. (2011). Item response modeling of forced-choice questionnaires. *Educational and Psychological Measurement, 71*(3):460–502.

Brown, A., and A. Maydeu-Olivares. (2013). How IRT can solve problems of ipsative data in forced-choice questionnaires. *Psychological Methods, 18*(1):36–52.

Campion, M.C., R.E. Ployhart, and W.I. MacKenzie. (2014). The state of research on situational judgment tests: A content analysis and directions for future research. *Human Performance, 27*:283–310.

Catano, V.M., A. Brochu, and C.D. Lamerson. (2012). Assessing the reliability of situational judgment tests used in high-stakes situations. *International Journal of Selection and Assessment, 20*(3):333–346.

Chan, D., and N. Schmitt. (1997). Video-based versus paper-and-pencil method of assessment in situational judgment tests: Subgroup differences in test performance and face validity perceptions. *Journal of Applied Psychology, 82*(1):143–159.

Chan, D., and N. Schmitt. (2002). Situational judgment and job performance. *Human Performance, 15*(3):233–254.

Christian, M.S., B.D. Edwards, and J.C. Bradley. (2010). Situational judgment tests: Constructs assessed and a meta-analysis of their criterion-related validities. *Personnel Psychology, 63*(1):83–117.

Clevenger, J., G.M. Pereira, D. Wiechmann, N. Schmitt, and V.S. Harvey. (2001). Incremental validity of situational judgment tests. *Journal of Applied Psychology, 86*:410–417.

Cortina, J.M. (1993). What is coefficient alpha? An examination of theory and applications. *Journal of Applied Psychology, 78*(1):98–104.

Crook, A.E., M.E. Beier, C.B. Cox, H.J. Kell, A.R. Hanks, and S.J. Motowidlo. (2011). Measuring relationships between personality, knowledge, and performance using single-response situational judgment tests. *International Journal of Selection and Assessment, 19*(4):364–373.

de la Torre, J., V. Ponsoda, I. Leenen, and P. Hontangas. (2012). *Examining the Viability of Recent Models for Forced-Choice Data*. Presented at the Meeting of the American Educational Research Association. Vancouver, British Columbia, Canada, April. Abstract available: http://www.aera.net/Publications/OnlinePaperRepository/AERAOnlinePaperRepository/tabid/12720/Owner/289609/Default.aspx [February 2015].

Edwards, J.R., and D.M. Cable. (2009). The value of congruence. *Journal of Applied Psychology, 94*:654–677.

Hanson, M.A., and W.C. Borman. (1995). *Development and Construct Validation of the Situational Judgment Test* (ARI Research Note 95-34). Alexandria, VA: U.S. Army Research Institute for the Behavioral and Social Sciences.

Hulin, C.L., F. Drasgow, and C.K. Parsons. (1983). *Item Response Theory: Applications to Psychological Measurement*. Homewood, IL: Dow Jones-Irwin.

Irvine, S.H., and P.C. Kyllonen, Eds. (2002). *Item Generation and Test Development*. Mahwah, NJ: Laurence Erlbaum Associates.

Jurafsky, D., and J.H. Martin. (2008). *Speech and Language Processing: An Introduction to Natural Language Processing, Computational Linguistics, and Speech Recognition*. Upper Saddle River, NJ: Prentice Hall.

Kolen, M.J., and R.L. Brennan. (2004). *Test Equating, Scaling, and Linking: Methods and Practices*. New York: Springer.

Krumm, S., F. Lievens, J. Hüffmeier, A.A. Lipnevich, H. Bendels, and G. Hertel. (2014). How "situational" is judgment in situational judgment tests? *Journal of Applied Psychology* (Epub). Available: http://www.researchgate.net/publication/264676090_How_Situational_Is_Judgment_in_Situational_Judgment_Tests [February 2015].

Kumar, E. (2011). *Natural Language Processing*. New Delhi, India: I.K. International Publishing House.

Lievens, F., and P.R. Sackett. (2006). Video-based versus written situational judgment tests: A comparison in terms of predictive validity. *Journal of Applied Psychology*, 91(5): 1,181–1,188.

Lievens, F., and P.R. Sackett. (2007). Situational judgment tests in high-stakes settings: Issues and strategies with generating alternate forms. *Journal of Applied Psychology*, 92(4):1,043–1,055.

Lievens, F., P.R. Sackett, and T. Buyse. (2009). The effects of response instructions on situational judgment test performance and validity in a high-stakes context. *Journal of Applied Psychology*, 94:1,095–1,101.

McDaniel, M.A., and N.T. Nguyen. (2001). Situational judgment tests: A review of practice and constructs assessed. *International Journal of Selection and Assessment*, 9:103–113.

McDaniel, M.A., and D.L. Whetzel. (2005). Situational judgment test research: Informing the debate on practical intelligence theory. *Intelligence*, 33(5):515–525.

McDaniel, M.A., F.P. Morgeson, E.B. Finnegan, M.A. Campion, and E.P. Braverman. (2001). Use of situational judgment tests to predict job performance: A clarification of the literature. *Journal of Applied Psychology*, 86:730–740.

McDaniel, M.A., N.S. Hartman, D.L. Whetzel, and W.L. Grubb, III. (2007). Situational judgment tests, response instructions, and validity: A meta-analysis. *Personnel Psychology*, 60(1):63–91.

McDonald, R.P. (1999). *Test Theory: A Unified Treatment*. Mahwah, NJ: Lawrence Erlbaum Associates.

Millsap, R.E., and J. Yun-Tein. (2004). Assessing factorial invariance in ordered-categorical measures. *Multivariate Behavioral Research*, 39(3):479–515.

Motowidlo, S.J., M.D. Dunnette, and G.W. Carter. (1990). An alternative selection procedure: The low-fidelity simulation. *Journal of Applied Psychology*, 75:640–647.

Motowidlo, S.J., M.P. Martin, and A.E. Crook. (2013). Relations between personality, knowledge, and behavior in professional service encounters. *Journal of Applied Social Psychology*, 43(9):1,851–1,861.

Mumford, T.V., F.P. Morgeson, C.H. Van Iddekinge, and M.A. Campion. (2008). The team role test: Development and validation of a team role knowledge situational judgment test. *Journal of Applied Psychology*, 93(2):250–267.

Nguyen, N.T., M.D. Biderman, and M.A. McDaniel. (2005). Effects of response instructions on faking a situational judgment test. *International Journal of Selection and Assessment*, 13(4):250–260.

Olson Buchanan, J.B., F. Drasgow, P.J. Moberg, A.D. Mead, P.A. Keenan, and M.A. Donovan. (1998). An interactive video assessment of conflict resolution skills. *Personnel Psychology*, 51(1):1–24.

Oswald, F.L., N. Schmitt, B.H. Kim, L.J. Ramsay, and M.A. Gillespie. (2004). Developing a biodata measure and situational judgment inventory as predictors of college student performance. *Journal of Applied Psychology*, 89(2):187–207.

Oswald, F.L., A.J. Friede, N. Schmitt, B.K. Kim, and L.J. Ramsay. (2005). Extending a practical method for developing alternate test forms using independent sets of items. *Organizational Research Methods*, 8(2):149–164.

Peeters, H., and F. Lievens. (2005). Situational judgment tests and their predictiveness of college students' success: The influence of faking. *Educational and Psychological Measurement*, 65(1):70–89.

Peterson, N.G., L.E. Anderson, J.L. Crafts, D.A. Smith, S.J. Motowidlo, R.L. Rosse, G.W. Waugh, R. McCloy, D.H. Reynolds, and M.R. Dela Rosa. (1999). *Expanding the Concept of Quality in Personnel: Final Report* (ARI Research Note 99-31). Alexandria, VA: U.S. Army Research Institute for the Behavioral and Social Sciences.

Ployhart, R.E. (2006). The predictor response process model. In J.A. Weekley and R.E. Ployhart, Eds., *Situational Judgment Tests: Theory, Measurement, and Application* (pp. 83–105). Mahwah, NJ: Lawrence Earlbaum Associates.

Ployhart, R.E., and M.G. Erhart. (2003). Be careful what you ask for: Effects of response instructions on the construct validity and reliability of situational judgment tests. *International Journal of Selection and Assessment, 11*(1):1–16.

Richman-Hirsch, W.L., J.B. Olson-Buchanan, and F. Drasgow. (2000). Examining the impact of administration medium on examinee perceptions and attitudes. *Journal of Applied Psychology, 85*:880–887.

Rumsey, M.G., and J.M. Arabian. (2014). Military enlistment selection and classification: Moving forward. *Military Psychology, 26*(3):221–251.

Smiderle, D., B.A. Perry, and S.F. Cronshaw. (1994). Evaluation of video-based assessment in transit operator selection. *Journal of Business and Psychology, 9*(1):3–22.

Tucker, J.S., A.N. Gesselman, and V. Johnson. (2010). *Assessing Leader Cognitive Skills with Situational Judgment Tests: Construct Validity Results* (Research Product 2010-04). Fort Benning, GA: U.S. Army Research Institute for the Behavioral and Social Sciences.

van der Linden, W.J. (1998). Optimal assembly of psychological and educational tests. *Applied Psychological Measurement, 22*(3):195–211.

van der Linden, W.J., and C.A.W. Glas, Eds. (2010). *Computerized Adaptive Testing: Theory and Practice*. New York: Kluwer Academic.

Wagner, R.K., and R.J. Sternberg. (1991). *Tacit Knowledge Inventory for Managers: User Manual*. San Antonio, TX: The Psychological Corporation.

Weekley, J.A., and C. Jones. (1997). Video-based situational testing. *Personnel Psychology, 50*(1):25–49.

Westring, A.J.F., F.L. Oswald, N. Schmitt, S. Drzakowski, A. Imus, B. Kim, and S. Shivpuri. (2009). Trait and situational variance in a situational judgment measure of goal orientation. *Human Performance, 22*(1):44–63.

Whetzel, D.L., and M.A. McDaniel. (2009). Situational judgment tests: An overview of current research. *Human Resource Management Review, 19*(3):188–202.

Whetzel, D.L., M.A. McDaniel, and N.T. Nguyen. (2008). Subgroup differences in situational judgment test performance: A meta-analysis. *Human Performance, 21*(3):291–309.

10

Assessment of Individual Differences Through Neuroscience Measures

Committee Conclusion: A wide variety of measures fall within the domain of neuroscience (e.g., direct neuroscience measures such as electroencephalography [EEG], positron emission tomography, magnetic resonance imagery [MRI], or functional MRI [fMRI] and indirect biomarkers of neural activity such as heart rate or eye blink). These measures may take multiple roles in the Army accession process including (a) monitoring test takers for constructs such as anxiety, attention, and motivation during other assessments; (b) use in research settings as criteria for evaluating other potential assessments; and (c) use as direct selection and classification assessments. Although the third role may be well in the future in terms of technically feasible and cost-effective assessment, the first two uses have near-term promise. The committee concludes that the neuroscience domain merits inclusion in a program of basic research with the long-term goal of improving the Army's enlisted accession system.

INTRODUCTION

The breadth, scope, and history of neuroscience are well summarized in the context of Army applications in the 2009 National Research Council report, *Opportunities in Neuroscience for Future Army Applications*. In that report, the Army was encouraged to engage in research and development "to take best advantage of variations in the neural bases of behavior that contribute to performance" (p. 4).

The current study committee was charged with recommending a research agenda based upon the biological basis of individual differences in behavior, including performance potential, and the determination of biomarkers indicating the state of biological systems that affect capability to perform. (See the complete Statement of Task in Chapter 1, Box 1-1; also see Appendix C for a more complete description of biomarkers.) Neuroscience methods may not be optimal for carrying out the job of testing recruits and classifying their performance capability—at least not at present. However, these methods do provide insight into how best to design test environments that allow recruits to perform up to their capacity, and they do have the potential to predict and even improve the learning of people in advance of test performance.

In this chapter, the committee uses the following measures, which it considers to incorporate aspects of neuroscience in the sense that they measure bodily functions that have been linked to psychological state or behavior. These measures include the following:

1. *psychophysiological measures,* such as heart rate and cardio rhythms, eye position, galvanic skin response, and pupil size;
2. *neuroimaging measures,* such as fMRI, electrical recording, and optical imaging with near infrared spectroscopy; and
3. *biochemical measures,* such as level of neuromodulators or hormones related to stress.

Appendix D provides a brief tutorial describing neural signals and their relationship to neuroimaging measures.

Most of the measures described above are not appropriate for routine use during large scale testing of recruits. It is often very difficult to interpret the results on an individual basis, and they can be influenced by many environmental factors. Moreover, their use with individuals may be seen as an invasion of privacy or could lead to even higher levels of stress than the testing itself. However, the committee believes there currently are less obtrusive methods, such as salivary measures of cortisol for monitoring stress during testing, that could benefit the Army's selection research and practice, and future neuroscience methodologies may become less intrusive and more effective in application. More generally, we suggest that neuroscience methods be used in research and field testing in which the measures would be used with volunteer samples of recruits to help determine the optimal methods of behavior testing given logistical and technical constraints. These neuroscience measures could currently be used to shape and improve the behavioral testing environment, thus reducing stress and increasing attention during testing. In the future they may allow better prediction of which individual will be best able to learn the material to be tested, and they may help in the design of tests or interventions that reduce impediments due to test anxiety, poor concentration, or other learning problems. We cover these possibilities below.

MONITORING TEST PERFORMANCE

For certain individuals, poor performance in a testing situation (especially timed high-stakes testing situations) can attribute to anxiety for reasons unrelated to the task, and several decades of theory and data support the idea that removal of off-task distraction (e.g., anxiety) through training can reduce interference with the primary task of test performance (Wine, 1971). More recent research has focused on performance failures at critical periods for lower-ability examinees (DeCaro et al., 2011). Although certain aspects of test anxiety can be appropriate due to the examinee's accurate expectation of not performing well on a test in a given domain (e.g., anxiety in a test or performance for which the examinee has insufficient skill or knowledge), the committee views inappropriate levels of test anxiety during assessments as a potential validity threat to the Armed Services Vocational Aptitude Battery (ASVAB) and the Tailored Adaptive Personality Assessment System (TAPAS; see Chapter 1 for further discussion of these tests).

As discussed in Chapter 6, the ability to perform under stress is an important attribute in many situations that require effective performance under situational pressures such as time, evaluation, or danger. To be successful in an Army career, performance in a wide range of stress and anxiety circumstances is necessary (Hancock and Szalma, 2008). The concept of a physiological reaction to stress—called strain—dates back to the idea of self-regulation producing homeostasis; that is, changes to the external environment—called stressors—result in physiological changes that maintain a body's internal environment within nominal operating parameters (Cannon, 1932). In this report, the committee focuses on measuring strain, the psychological consequences of stress. Psychological strain can be objectively defined as any action or situation that causes acute release of stress hormones, predominantly catecholamines and glucocorticoids.[1]

An important difference exists between stress response and emotional response. Stress is the general application of an external change, and the stress response is the resulting release of hormones to cause some internal compensating change. Emotional responses of fear and anxiety are strains related to the fight or flight response (see also Chapter 6) with an increased state of arousal in the autonomic nervous system—for example, cardiac, respiratory, and perspiration changes. Stress can induce such emotions, and

---

[1] Catecholamines in the human body include epinephrine, norepinephrine, and dopamine. The most well-known glucocorticoid is cortisol, which is necessary for cardiovascular, metabolic, immunologic, and homeostatic functions.

some emotions in turn can cause stress. In the research literature, four basic characteristics of situations are emerging as common triggers for stress response in humans: novelty, unpredictability, threats to survival or ego, and low sense of control.

Anxiety, for example, is an emotional response to a stress that is often longer lasting than the duration of the stress that triggers it. The trigger stress can sometimes be difficult to identify or predict within aspects of a given situation (simple or complex; they are highly subjective), and therefore it can be difficult for people to self-manage anxiety or for situations to be modified to manage anxiety. Increases in anxiety tend to result in changes to the autonomic nervous system and to brain activity (van Stegeren, 2009). Stress induces changes to brain activity patterns, commonly in the anterior cingulate cortex and orbitofrontal regions (Cannon, 1932; Dedovic et al., 2009; Lupien, 2009; van Stegeren, 2009). Stress and anxiety are normal parts of human homeostatic reaction that is usefully incorporated into a model of adaptability (see Figure 10-1) historically in-

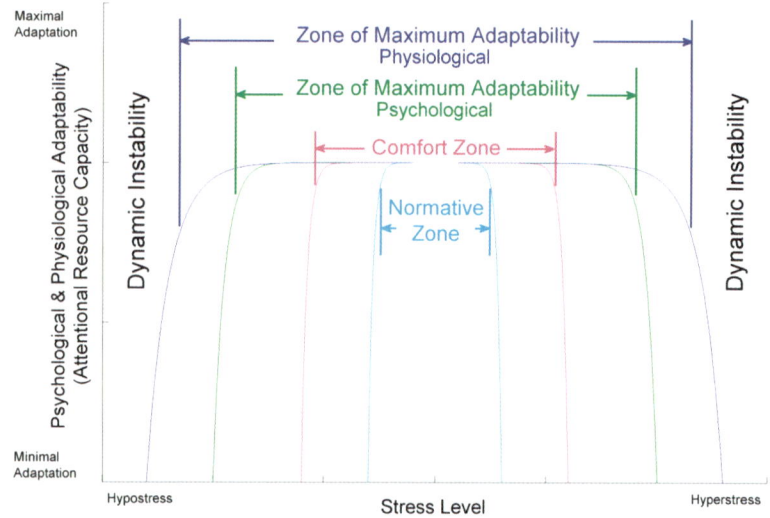

FIGURE 10-1 Physiological adaptive capability (blue line) and psychological adaptive capability (green line) can be equated with attentional resource capacity, plotted qualitatively here as a function of stress level. An individual's performance will degrade at either end of the curve. Homeostatic pressure directs mental and physiological processes to move toward the center of the stress axis. There are also a central normative zone and a slightly more flexible comfort zone.
SOURCE: Hancock, P.A., and J.S. Warm. (1989). A dynamic model of stress and sustained attention. *Human Factors*, (31)5:528. Copyright SAGE Publications. ©Reproduced by permission of SAGE Publications.

spired by Yerkes-Dodson and proposed by Hancock (Hancock and Warm, 1989). In this model, ultralow or ultrahigh levels of anxiety or stress are outside the window of an individual's ability to adapt (see also Chapter 7 for a discussion of adaptability).

It is possible to monitor levels of stress through the analysis of saliva for cortisol hormone levels (Hucklebridge et al., 1998). Testing situations can pose a cognitive challenge to some individuals that then lead to elevated hormonal levels within a few minutes of testing. By taking saliva samples before and after the test, one can examine the effect of stress level on test performance. Moreover, as described below, there are effective means for combatting the stress produced by cognitive challenges (Fan et al., 2014). These could be used with recruits who demonstrate an unusually strong response to the test situation, perhaps focusing on those whose performance appears out of line with predictions or other knowledge about the person.

DESIGNING BETTER TESTS AND TESTING ENVIRONMENTS

Comparing Environments

A testing situation is an inherently evaluative experience and therefore often a stressful one. However, both the environment in which the test is conducted and the format of the test itself can contribute to increased stress and anxiety. As a part of a quasi-experimental research effort, neuroscience measures such as salivary cortisol could be used to assess examinees in test environments that vary naturally. For example, environments that are hot or noisy, have poor ventilation, or require longer testing periods may be more stressful than others. Environments that appear similar may still lead to different stress responses because of the context in which stress level is measured, such as the time of day, the test taker's preceding activities, or the presence of sleep deprivation.

In addition to the stress response, sympathetic and parasympathetic activity can be measured by use of heart rate variability measures (Axelrod et al., 1981). Scalp electrodes can be used to assess the presence of signs of attentive processing during the session (Posner, 2008).

Comparing Tests

How does cognitive activity vary as a function of different types of test content? One can measure the time to process information in the test by examining successive eye movements and dwell time on relevant information (Rayner and Pollatsek, 1989). To get a finer picture of the speed and location of brain activity in processing the test material, one can combine

EEG and neuroimaging methods to obtain the location and timing of brain activity, as has been done in the study of word and sentence processing (Posner et al., 1999). This information might enable improvements in the presentation and content of the test material used for recruits.

PREDICTING HUMAN PERFORMANCE

This section summarizes research supporting the possibility of predicting how recruits may perform in testing situations through neuroscience methods and also of modifying (mitigating) their stress reactions to testing. Although the committee cites evidence for each of these points, they remain areas of ongoing research that will require additional effort to develop the fundamental understanding necessary for long term potential application in assessment processes. However, the preliminary evidence offers great promise for the future application of neuroscience methods in improving predictability and performance by military personnel.

Brain Changes in Learning

Learning modifies the connectivity and activity of the brain. If, for example, one is presented with a word (e.g., "hammer") and then is asked to develop a use for it (e.g., pounding) a set of neural areas become active. These include areas in the left frontal and temporal lobe, the anterior cingulate and portions of the right cerebellum. However, if one practices a few minutes with the same association, that pathway changes: the frontal, cingulate, and cerebellar activations diminish and the visual and motor brain areas that resemble those used in reading aloud remain (Raichle et al., 1994).

This finding is a good example of what happens when carrying out any task. There are a set of brain areas that are active, and they are orchestrated over a short period to produce behavior (Posner and Raichle, 1994). Practice changes the areas of activation, concentrates them, and improves the connections between them (Zatorre et al., 2012). These changes are reflected behaviorally in terms of a more predictable, proceduralized, and faster response.

Predicting Behavior from Brain Activity

Recently several forms of fMRI have been employed to predict performance in cognitive tasks. For example, in one study, resting-state fMRI was used to assess the functional connectivity between brain areas that were related to a visual discrimination task to be performed subsequently by human participants (Baldassarre et al., 2012). The study found that resting-state connectivity predicted both initial performance and speed of learning in the subsequent task. A similar study in the motor system measured connectivity with high-density EEG and was able to predict the rate of subsequent motor learning (Wu et al., 2014). So far, only rather simple cognitive tasks have been predicted from the connectivity pattern. However, connectivity in brain areas related to executive attention and working memory (see Chapter 2) may have a robust influence on many tasks that are critically important in military situations, such as map reading, navigation, and decision making. Research into prediction of such high-level skills from patterns of connection derived from resting state MRI and other neuroscience methods would be of high priority.

Many studies (See Zatorre et al., 2012, for a summary) show that the act of learning a skill improves white matter connectivity between brain areas related to the task. It seems reasonable that assessment of the preexisting connectivity should predict the rate of improvement.

Although the studies above use tasks where the areas of activation and patterns of connectivity are known in advance, new statistical methods suggest it will be possible in the future to make predictions even in situations where there is little advanced knowledge of the brain areas and connectivity involved (Friston et al., 2003; Norman et al., 2006).

Another possible route to prediction is via measures of the neural systems and genetics related to individual differences in personality (Canli et al., 2001; DeYoung and Gray, 2009). It has been common to view differences in personality as enduring aspects that describe traits that can influence behavior (Goldberg and Rosolack, 1994). Efforts have been made to describe the neural systems involved and to understand the genetic basis of some of these traits (Canli and Lesch, 2007). This is a kind of construct validation process. One of the most important personality traits is conscientiousness, which is related to effort control in the research literature addressing the temperament of children (Rothbart, 2011). Measures of effortful control assessed in childhood have proven predictive of a wide range of behaviors and outcomes over the life span (Moffitt et al., 2011). Neural systems related to effortful control involve a network including the anterior cingulate, anterior insula, and underlying striatum (Posner et al., 2007). Some dopaminergic genes that influence the efficiency of this network have been studied. Research into effortful control suggests that higher effortful control relates to more functional adult behaviors, including the ability to learn new things (Posner and Rothbart, 2007) and emotion regulation (see Chapter 6). Thus, studies of the neural systems related to effortful control would be of great value in the evaluation of recruits.

Training

In addition to predicting which recruits will do better in subsequent tests, neuroscience measures have been important in evaluating two methods that might reduce stress and thereby potentially improve the ability to learn new material.

Training working memory is one method that has led to improvements in activation of lateral areas of the frontal and parietal lobes (Olesen et al., 2004). In initial studies, it was reported to transfer to general intelligence and other cognitive tasks (Jaeggi et al., 2008), although transfer to intelligence and other measures remote from the training remains a highly disputed issue, given subsequent data failing to find such empirical support (Redick et al., 2013).

Another method with potential for wide-ranging consequences for stress reduction as well as cognitive performance involves mindfulness meditation training (Tang and Posner, 2014). In one study, 1- to 4-week mindfulness training reduced stress to a cognitive challenge as measured by salivary cortisol secretion in a dose-dependent fashion (Fan et al., 2014). In other studies, 1 to 4 weeks of mindfulness meditation in comparison to a control group given relaxation training improved attention and mood and changed white matter connectivity between the anterior cingulate and other areas as measured by diffusion tensor imaging (Tang et al., 2012). As mentioned above, such white matter changes can lead to improvements in learning.

Given these two key methods and the promise for others to arise, the committee suggests further research designed to use neuroscience methods to improve the testing experience and training of recruits.

RESEARCH RECOMMENDATION

The U.S. Army Research Institute for the Behavioral and Social Sciences should pursue a program for investigating the potential for robust and objective neurophysiological biomarkers that can serve to refine and augment assessments currently in use or under development for future utilization. These biomarkers may include, among others, eye tracking, physiological reactions (galvanic skin response, cardio rhythms, etc.), medium term endocrine measures (cortisol, neurochemical markers), brain activity measures, and static and functional brain imaging. This program investigating neurophysiological biomarkers should prepare to address challenges in both what to measure and how to accomplish the measurements technically, first in the laboratory setting and eventually in field settings. The program should support research in relevant biomarker development for use in the following roles:

A. Research seeking refinement of current and future Army assessments (e.g., the Armed Services Vocational Aptitude Battery and Tailored Adaptive Personality Assessment System) through a deeper understanding of the constructs measured in selection and classification testing. In this role, biomarkers may reveal underlying neurophysiological correlates of constructs of interest (e.g., cortisol as a biomarker for anxiety). Deeper understanding of physiology has the potential to differentiate complex constructs or alternatively to reveal the relative strength of measures.
B. Independent use of biomarkers as direct selection instruments or their use in combination with traditional assessments. Research should identify biomarker correlates (e.g., consistent gaze, pupilometry, reaction time in a vigilance test) of abilities and outcomes. Test stimuli or conditions for eliciting biological responses from test takers (e.g., simulated rifle drill or other novel muscle coordination task to assess parietal-dominant brain) should be developed.

In addition to these key roles, there might be other ways biomarker development could contribute to a selection and classification program:

A. Monitoring candidates for attributes such as anxiety during assessment and offering training on mitigation strategies for applicants not selected on the basis of their test scores. Such attributes can contribute to bias in test scores, and success in controlling for the effects of these attributes can result in more valid assessment. The committee expects challenges in determining whether applicants' observed performance reflects their true ability (e.g., whether applicants are experiencing normal performance stress or an interfering level of anxiety). Additionally, we expect challenges in designing a simple and effective mitigation program.
B. Basic research to apply modern neurophysiological tools to model test-taker response data (e.g., response time distributions, answer patterns that may suggest unmotivated responding or intentional distortion).

REFERENCES

Axelrod, S., D. Gordon, F.A. Ubel, D.C. Shannon, A.C. Barger, and R.J. Cohen. (1981). Power spectrum analysis of heart rate fluctuation: A quantitative probe of beat-to-beat cardiovascular control. *Science*, 213:220–222.

Baldassarre, A., C.M. Lewis, G. Committeri, A.Z. Snyder, G.L. Romani, and M. Corbetta. (2012). Individual variability in functional connectivity predicts performance of a perceptual task. *Proceedings of the National Academy of Sciences of the United States of America*, 109(9):3,516–3,521.

Canli, T., and K.P. Lesch. (2007). Long story short: The serotonin transporter in emotion regulation and social cognition. *Nature Neuroscience, 10*:1,103–1,109.

Canli, T., Z. Zhao, J.E. Desmond, E. Kang, J. Gross, and J.D.E. Gabrieli. (2001). An fMRI study of personality influences on brain reactivity to emotional stimuli. *Behavioral Neuroscience, 115*:33–42.

Cannon, W.B. (1932). *The Wisdom of the Body*. New York: W.W. Norton.

DeCaro, M.S., R.D. Thomas, N.B. Albert, and S.L. Beilock. (2011). Choking under pressure: Multiple routes to skill failure. *Journal of Experimental Psychology General, 140*(3):390–406.

Dedovic, K., A. Duchesne, J. Andrews, V. Engert, and J.C. Pruessner. (2009). The brain and the stress axis: The neural correlates of cortisol regulation in response to stress. *Neuroimage, 47*(3):864–871.

DeYoung, C.G., and J.R. Gray. (2009). Personality neuroscience: Explaining individual differences in affect, behavior, and cognition. In P.J. Corr and G. Matthews, Eds., *Cambridge Handbook of Personality Psychology* (pp. 323–346). New York: Cambridge University Press.

Fan, Y.X., Y.Y. Tang, and M.I. Posner. (2014). Cortisol level modulated by integrative meditation in a dose dependent fashion. *Stress and Health, 30*(1):65–70.

Friston, K.J., L. Harrison, and W. Penny. (2003) Dynamic causal modelling. *Neuroimage, 19*(4):1,273–1,302.

Goldberg, L., and T.K. Rosolack. (1994). The Big Five factor structure as an integrative framework. In C.F. Haverson, G.A. Kohnstamm, and R. Martin, Eds., *The Developing Structure of Temperament and Personality from Infancy to Adulthood* (pp. 7–35). Hillsdale, NJ: Lawrence Erlbaum Associates.

Hancock, P.A., and J.L. Szalma. (2008). *Performance Under Stress*. Aldershot, England; Burlington, VT: Ashgate.

Hancock, P.A., and J.S. Warm. (1989). A dynamic model of stress and sustained attention. *Human Factors, 31*(5):519–537.

Hucklebridge, F., A. Clow, and P. Evans. (1998). The relationship between salivary secretory immunoglobulin A and cortisol: Neuroendocrine response to awakening and the diurnal cycle. *International Journal of Psychophysiology, 31*:69–76.

Jaeggi, S.M., M. Buschkuehl, J. Jonides, and W.J. Perrig. (2008). Improving fluid intelligence with training on working memory. *Proceedings of the National Academy of Sciences of the United States of America, 105*(19):6,829–6,833.

Lupien, S.J. (2009). Brains under stress. *Canadian Journal of Psychiatry, 54*(1):4–5.

Moffitt, T.E., L. Arseneault, D. Belsky, N. Dickson, R.J. Hancox, H.L. Harrington, R. Houts, R. Poulton, B.W. Roberts, S. Ross, M.R. Sears, W.M. Thomson, and A. Caspi. (2011). A gradient of childhood self-control predicts health, wealth and public safety. *Proceedings of the National Academy of Sciences of the United States of America, 108*(7):2,693–2,698.

National Research Council. (2009). *Opportunities in Neuroscience for Future Army Applications*. Committee on Opportunities in Neuroscience for Future Army Applications. Board on Army Science and Technology, Division on Engineering and Physical Sciences. Washington, DC: The National Academies Press.

Norman, K.A., S.M. Polyn, G.J. Detre, and J.V. Haxby. (2006). Beyond mind-reading: Multivoxel pattern analysis of fMRI data. *Trends in Cognitive Science, 10*(9):424–430.

Olesen, P.J., H. Westerberg, and T. Klingberg. (2004). Increased prefrontal and parietal activity after training of working memory. *Nature Neuroscience, 7*:75–79.

Posner, M.I. (2008). Measuring alertness. In D.W Pfaff and B.L. Kieffer, Eds., *Molecular and Biophysical Mechanisms of Arousal, Alertness and Attention* (pp. 193–199). Boston, MA: Wiley Blackwell.

Posner, M.I., and M.E. Raichle. (1994). *Images of Mind*. Scientific American Library. New York: W.H. Freeman.

Posner, M.I., and M.K. Rothbart. (2007). Research on attention networks as a model for the integration of psychological science. *Annual Review of Psychology, 58*:1–23.

Posner, M.I., Y. Abdullaev, B.D. McCandliss, and S.E. Sereno. (1999). Anatomy, circuitry, and plasticity of word reading. In J. Everatt, Ed., *Reading and Dyslexia: Visual and Attentional Processes* (pp. 137–162). London: Routledge.

Posner, M.I., M.K. Rothbart, and B.E. Sheese. (2007). Attention genes. *Developmental Science, 10*:24–29.

Raichle, M.E., J.A. Fiez, T.O. Videen, A.M.K. McLeod, J.V. Pardo, P.T. Fox, and S.E. Petersen. (1994). Practice related changes in the human brain: Functional anatomy during nonmotor learning. *Cerebral Cortex, 4*:8–26.

Rayner, K., and A. Pollatsek. (1989). *Psychology of Reading*. Englewood Cliffs, NJ: Prentice-Hall.

Redick, T.S., Z. Shipstead, T.L. Harrison, K.L. Hicks, D.E. Fried, D.Z. Hambrick, M.J. Kane, and R.W. Engle. (2013). No evidence of intelligence improvement after working memory training: A randomized, placebo-controlled study. *Journal of Experimental Psychology General, 142*:359–379.

Rothbart, M.K. (2011). *Becoming Who We Are: Temperament, Personality, and Development*. New York: Guilford Press.

Tang, Y.Y., and M.I. Posner. (2014). Training brain networks and states. *Trends in Cognitive Sciences, 18*(7):345–350.

Tang, Y.Y., Q. Lu, M. Fan, Y. Yang, and M.I. Posner. (2012). Mechanisms of white matter changes induced by meditation. *Proceedings of the National Academy of Sciences of the United States of America, 109*(26):10,570–10,574.

van Stegeren, A.H. (2009). Imaging stress effects on memory: A review of neuroimaging studies. *Canadian Journal of Psychiatry, 54*(1):16–27.

Wine, J. (1971). Test anxiety and direction of attention. *Psychological Bulletin, 76*(2):92–104.

Wu, J., R. Srinivasan, A. Kaur, and S.C. Cramer. (2014). Resting-state cortical connectivity predicts motor skill acquisition. *Neuroimage, 91*:84–90.

Zatorre, R.J., R.D. Fields, and H. Johansen-Berg. (2012). Plasticity in gray and white; Neuroimaging changes in brain structure during learning. *Nature Neuroscience, 15*:528–536.

Section 6

The Research Agenda

11

The Research Agenda

IMPLEMENTATION

The charge to the committee requested the recommendation of a basic research agenda, for implementation by the U.S. Army Research Institute for the Behavioral and Social Sciences (ARI), that might in time lead to improvements in the Army enlisted soldier selection process. In developing the recommended research agenda and considering an implementation strategy that includes the necessary funding level, the committee excluded possible methods of improving selection that were, in the committee's judgment, beyond the basic research stage. However, the committee recognizes that aspects of the research topics identified in this report are already under investigation by ARI and other entities to varying degrees, while other aspects may be in the process of being developed and implemented.

As described in the report's first chapter, throughout its work the committee recognized the importance of developing selection systems based on criteria of interest to organizational values. However, the Army's currently used selection tools and systems apply to multiple missions, environments, and criteria that represent its organizational values, and the Army is also forward-looking in considering jobs, environments, and selection in the future. Thus the committee was instructed to think broadly about the selection of military personnel across all occupational specialties rather than to consider selection issues that might be unique to any specific outcome or function. The research recommendations, as compiled and restated in this chapter, reflect the committee's requirement that a conceptual or empirical link could be identified between an attribute under consideration and one or more outcomes that constitute a component of overall individual or team effectiveness. (The reader is referred back to Table 1-1 in Chapter 1 for the grid presenting the links between the research domains included in this report and many of the outcomes identified by the committee as potentially of importance to the Army.)

In considering the implementation of the research agenda, the question of the necessary funding level for future ARI basic research is of key import. Because the committee recognized that it lacked critical expertise and insight into the Army organization and missions, this report was developed on the basis that the Army would need to identify the outcomes strategically of greatest value to its mission(s), then basic research domains linked to those outcomes would become higher in priority. Funding allocations would be impacted by such a priority scheme.

Absent priorities assigned to the 10 substantive recommendations made in this report, the committee sees each of the areas as independently worthy of pursuit. The research topics have been grouped into relevant sections in the report, based upon the taxonomic system described in Chapter 1, and interrelated topics could be developed into integrative research programs. However, to produce findings that have the potential to improve the quality of Army selection decisions in the relatively near term, the committee believes all topics identified in this report should be pursued at levels commensurate with the outcomes of greatest import to the Army.

If all research topics could be pursued, a modest start would be to fund one project in each of these 10 areas. A reasonable average funding level for these projects might be $350,000 per year. We note that this funding level is consistent with the typical current funding level for basic research projects supported from ARI's Personnel Performance and Training budget line. This funding would be exclusive to the basic research program and would not include formal validity studies or applied programs of research prior to implementation. Note that the per-project funding cited above is an average value; work in some domains can be expected to be more costly than in others, and different research strategies within a domain may be more costly than others. Equipment needs and participant payment costs are among the features that are likely to vary across domains and across projects.

Thus, a research budget of $3.5 million would support this initial plan of one project per substantive area per year. One project certainly reflects progress. But each substantive research domain is multifaceted, and multiple projects per area would permit quicker progress and potential synergies across projects. So a more ambitious plan would be to fund two projects per year in each of these 10 areas, thus suggesting a research budget of $7 million. To be clear, this represents funding for basic research. Follow-up research moving toward operational use of new measures (e.g., field validation studies) will be necessary but is beyond the committee's charge.

In the committee's opinion, to implement such a program effectively and expeditiously would require a funding commitment in the range of $3.5 million per year (supporting one project per substantive area) to $7 million per year (supporting two projects per substantive area) in order to support research on potential enhancement of enlisted soldier selection.

RESEARCH TOPICS: COMMITTEE CONCLUSIONS AND RECOMMENDATIONS

For the convenience of the reader, this section of the report's final chapter restates the conclusions and recommendations that were originally presented in each of the relevant research topic chapters (Chapters 2 through 10) and that, combined, make up the committee's recommended research agenda for ARI to take its basic research program to the next leap forward in identifying, assessing, and assigning quality personnel.

Fluid Intelligence, Working Memory Capacity, Executive Attention, and Inhibitory Control (Chapter 2)

Committee Conclusion

The constructs of fluid intelligence (novel reasoning), working memory capacity, executive attention, and inhibitory control are important to a wide range of situations relevant to the military, from initial selection, selection for a particular job, and training regimes to issues having to do with emotional, behavioral, and impulse control in individuals after accession. These constructs reflect a range of cognitive, personality, and physiological dimensions that are largely unused in current assessment regimes. The committee concludes that these topics merit inclusion in a program of basic research with the long-term goal of improving the Army's enlisted accession system.

Research Recommendation: Fluid Intelligence, Working Memory Capacity, and Executive Attention

> The U.S. Army Research Institute for the Behavioral and Social Sciences should support research to understand the psychological, cognitive, and neurobiological mechanisms underlying the constructs of fluid intelligence (novel reasoning), working memory capacity, and executive attention.

A. Research should be conducted to ascertain whether these constructs reflect a common mechanism or are highly related but distinct mechanisms.
B. Assessments reflecting the results of research into the commonality versus distinctness of these constructs should be developed for purposes of validity investigations.
C. Ultimately, the basic research results from items A and B above should be used to inform research into time-efficient, computer-automated assessment(s).

Research Recommendation: Inhibitory Control

The U.S. Army Research Institute for the Behavioral and Social Sciences should support research to further understanding of inhibitory control, including but not limited to the following lines of inquiry:

A. Develop time-efficient, computer-automated self-report and behavioral assessments of inhibitory control capacity that demonstrate convergence with neurophysiological indices, as well as differentiation from constructs considered distinct from inhibitory control.
B. Examine the extent to which inhibitory control—as assessed through self-report, task-behavioral, and physiological response measures—predicts performance outcomes of interest (e.g., accidents, disciplinary incidents) and understand the common and unique aspects of the different assessment approaches in terms of underlying processes tapped by each and how these processes relate to performance.

Cognitive Biases (Chapter 3)

Committee Conclusion

Cognitive biases, such as confirmation bias, anchoring, overconfidence, sunk cost, availability, and others, appear broadly relevant to the military because of findings, from both the analysis of large-scale disasters and the broader literature on cognitive biases, that show how irrational decision making results from failing to reflect on choices. Research on a tendency to engage in cognitive biases as a stable individual-differences measure is limited, and there are measurement challenges that must be dealt with before operational cognitive bias assessment could be implemented. The conceptual relevance of this topic, paired with the limited research to date, which takes an individual-differences orientation, leads the committee to conclude that cognitive biases merit inclusion in a program of basic research with the long-term goal of improving the Army's enlisted accession system.

Research Recommendation

The U.S. Army Research Institute for the Behavioral and Social Sciences should support research to understand cognitive biases and heuristics, including but not limited to the following topics:

A. Research should be conducted to ascertain whether various cognitive biases and heuristics are accounted for by common bias susceptibility factors or whether various biases reflect distinct constructs (e.g., confirmation bias, fundamental attribution error).
B. A battery of cognitive bias and heuristics assessments should be developed for purposes of validity investigations.
C. Research should be conducted to examine the cognitive, personality, and experiential correlates of susceptibility to cognitive biases. This should include both traditional measures of personality and cognitive abilities (e.g., the Armed Services Vocational Aptitude Battery), and information-processing measures of factors such as working memory, executive attention, and inhibitory control.
D. Research should be conducted to identify contextual factors, that is, situations in which cognitive biases and heuristics may affect thought and action, and then to develop measures of performance in such situations, for use as criteria in studies aimed at understanding how cognitive biases affect performance. The research should consider the differentiating characteristics of contexts that determine when the use of heuristics for "fast and frugal" decision making might be beneficial, and when such thinking is better thought of as biased and resulting in poor decision making.

Spatial Abilities (Chapter 4)

Committee Conclusion

A spatial ability measure, Assembling Objects (AO), is included in the Armed Services Vocational Aptitude Battery (ASVAB). Research suggests incremental validity for spatial measures over general mental ability measures in predicting important military outcomes. Research also suggests that sex differences vary across different operationalizations of spatial ability. Together, these findings suggest exploring varying approaches to the measurement of spatial abilities to ascertain whether the AO test is the best measure of spatial ability for military selection and classification. The committee concludes that spatial ability merits inclusion in a program of basic research with the long-term goal of improving the Army's enlisted accession system.

Research Recommendation

> The U.S. Army Research Institute for the Behavioral and Social Sciences should support research to understand facets and assessment methods in the domain of spatial abilities, including the following research lines of inquiry:
>
> A. Identify or develop measures of various facets of spatial ability, with particular attention to the role of technology to overcome prior limitations in test-item formats.
> B. Examine the interrelationships among various facets of spatial ability, including but not limited to spatial relations, spatial orientation, and spatial visualization.
> C. Examine sex differences on the various facets of spatial ability, as well as the degree to which sex differences are mitigated or accentuated by various forms of training on the facets of spatial ability.
> D. Develop measures reflecting various work outcomes that can be used as criterion measures in evaluating the validity of various measures of spatial ability.

Teamwork Behavior (Chapter 5)

Committee Conclusion

Research has identified a number of individual-differences attributes that are broadly predictive of success in a team environment. There has also been progress in identifying attributes that when aggregated across team members (e.g., mean level of cognitive ability, minimum agreeableness), are predictive of team effectiveness. More research is needed to expand and amplify this work in the context of potential utility in military accession. The committee concludes that the teamwork knowledge, skills, abilities, and other characteristics (KSAO) domain merits inclusion in a program of basic research with the long-term goal of improving the Army's enlisted accession system.

Research Recommendation

> The U.S. Army Research Institute for the Behavioral and Social Sciences should support research on individual- and team-level knowledge,

skills, abilities, and other characteristics that influence the collective capacity to perform. Future research should include the following objectives:

A. Develop a better understanding of, and new metrics to operationalize, team outcomes and effectiveness. In addition, new technologies should be explored to better assess teamwork behaviors beyond paper-and-pencil measures.
B. Identify individual and team cognitions, affect/motivation, and behaviors that are linked to successful team outcomes and effectiveness. Essential to this is developing methods of team task analysis.
C. Identify optimal within-individual profiles that are linked to team effectiveness. This research should also consider types of team structures, tasks, and environmental conditions that moderate relationships between profile attributes and their combined influence on team processes and outcomes.
D. Investigate the effects of teamwork training and team experiences on the predictive power of individual-differences measures.

Hot Cognition: Defensive Reactivity, Emotional Regulation, and Performance under Stress (Chapter 6)

Committee Conclusion

"Hot cognition" includes the topics of defensive reactivity, emotional regulation, and performance under stress. Research and military experience suggest that the ability to perform well in situations that elicit emotional responses is important in many contexts that are relevant to the military. Research on performance has tended to underplay the role emotions can play in governing behavior, whether for good or bad. The committee concludes that the hot cognition domain merits inclusion in a program of basic research with the long-term goal of improving the Army's enlisted accession system.

Research Recommendation

The U.S. Army Research Institute for the Behavioral and Social Sciences should support research to understand issues in the domain of hot cognition:

A. Research should explore behavioral performance measures and also physiological measures of dispositional defensive reactivity, such as the eye-blink startle measure and other biological indicators (biomarkers) of fear activation, and more generally other traits conceived as "biobehavioral." Research should examine how biobehavioral dispositions like defensive reactivity relate to and are distinct from other personality constructs such as the Big Five (Openness, Conscientiousness, Extraversion, Agreeableness, and Neuroticism). In addition, research should compare the predictive validity of trait dispositions as assessed by physiological or behavioral measures in relation to survey assessments and examine how traits affect performance outcomes in differing situational contexts (e.g., impact of dispositional boldness on behavioral effectiveness in social versus affective versus workplace versus battlefield context).

B. Research should clarify how emotions and cognitions together affect human capability and performance and should expand understanding of the physiological bases for emotional regulation. Key themes include neural mechanisms of inhibition, the role of the prefrontal cortex in higher cognitive control including affective processing, and the role of the dorsal region of the anterior cingulate cortex in monitoring conflicts (e.g., conflict between emotional and cognitive influences on moral dilemma tasks).

C. Research should explore measuring emotional regulation with established forms of assessment such as rating scales, situational judgment tests, and performance measures (e.g., delay-of-gratification measures, emotional conflict tests, cooperation versus competition tasks).

D. Research should examine the conditions that improve or diminish cognition and performance under stress, in order to develop measures of susceptibility to stress.

E. Research should evaluate whether susceptibility to stress is contingent on the type of stressor (e.g., time pressure, peer pressure, fatigue) and whether there are cognitive, personality, and experiential correlates of susceptibility.

Adaptability and Inventiveness (Chapter 7)

Committee Conclusion

The military has a strong interest in adaptive behavior, expressed in terms of assessing novel problems and solving them or acting upon them effectively. Research indicates two promising lines of inquiry. The first would use measures of frequency and quality of ideas generated in open-ended tasks, which have demonstrated incremental validity over and above measures of general cognitive ability for predicting important outcomes related to work performance. The second line of inquiry would use narrow personality constructs to predict adaptive behavior and inventive/creative problem solving. Thus, the committee concludes that idea generation measures and narrow personality measures specific to adaptability and inventiveness merit inclusion in a program of basic research with the long-term goal of improving the Army's enlisted accession system.

Research Recommendation

> The U.S. Army Research Institute for the Behavioral and Social Sciences should support research to understand constructs and assessment methods in the domains of adaptability/inventiveness and adaptive performance, including but not limited to the following topics:
>
> A. Compare alternative approaches to the measurement and scoring of idea generation as a cognitive measure of adaptability/inventiveness.
> B. Use existing literature, theory, and empirical research to identify and develop narrow personality measures as candidates for predicting adaptive performance.
> C. Develop a range of measures of relevant work criteria that reflect adaptive performance in research studies.
> D. Examine the use of these personality and idea generation measures in predicting the above adaptive performance criteria.

Psychometrics and Technology (Chapter 8)

Committee Conclusion

The military has long been in the forefront of modernized operational adaptive testing. Recent research offers promise for improvements in measurement in a variety of areas, including the application and modeling of forced-choice measurement methods; development of serious gaming; and pursuing Multidimensional Item Response Theory (MIRT), Big Data analytics, and other modern statistical tools for estimating applicant standing on attributes of interest with greater efficiency. Efficiency is a key issue, as the wide range of substantive topics recommended for research in this report may result in proposed additions to the current battery of measures administered for accession purposes. The committee concludes that such advances in measurement and statistical models merit inclusion in a program of basic research with the long-term goal of improving the Army's enlisted accession system.

Research Recommendation

Modern measurement methods come with the promise of increasing precision, validity, efficiency, and security of current, emerging, and future forms of assessment. The U.S. Army Research Institute for the Behavioral and Social Sciences should continue to support developments to advance psychometric methods and data analytics.

A. Potential topics of research on Item Response Theory (IRT) include the use of multidimensional IRT models, the application of rank and preference methods, and the estimation of applicant standing on the attributes of interest with greater efficiency (e.g., via automatic item generation, automated test assembly, detecting item pool compromise, multidimensional test equating, using background information in trait estimation).
B. Ecological momentary assessments (e.g., experience sampling) and dynamic interactive assessments (e.g., team interaction, gaming, and simulation) yield vast amounts of examinee data, and future research should explore the new challenges and opportunities for innovation in psychometric and Big Data analytics.
C. Big Data analytics also may play an increasingly important role as candidate data from multiple diverse sources becomes increasingly available. Big Data methods designed to find structure in datasets with many more columns (variables) than rows (candidates) might help identify robust variables, important new constructs, interactions between constructs, and nonlinear relationships between those constructs and candidate outcomes.

Situations and Situational Judgment Tests (Chapter 9)

Committee Conclusion

The ability to use judgment to interpret, evaluate, and weigh alternate courses of action appropriately and effectively is relevant to a wide variety of situations within the military. Various streams of research, including new conceptual and measurement developments in assessing situational judgment, as well as evidence of consistent incremental validity of situational judgment measures over cognitive ability and personality measures for predicting performance in various work settings, lead the committee to conclude that measures of situational judgment merit inclusion in a program of basic research with the long-term goal of improving the Army's enlisted accession system.

Research Recommendation

The U.S. Army Research Institute for the Behavioral and Social Sciences should support research to understand constructs and assessment methods specific to the domain of situational judgment, including but not limited to the following lines of inquiry:

A. Develop situational judgment tests with items reflecting constructs that are otherwise difficult to assess using other tests, that are important, and that show promise for validity (e.g., prosocial knowledge, team effectiveness).
B. Consider innovative formats for presenting situations (e.g., ranging from simple text-based scenarios to dynamic and immersive computer-generated graphics), capturing examinee responses (e.g., open-ended, voice, gestures, facial expressions, eye movements, reaction times), and evaluating examinee responses (e.g., advanced natural language processing, automated reasoning, machine learning).
C. Develop and explore psychometric models and methods that can accommodate the rich array of data that innovative assessment methods for situational judgment may yield, facilitating the development of psychometrically and practically equivalent assessments, and improving reliability and testing efficiency.

Assessment of Individual Differences Through Neuroscience Measures (Chapter 10)

Committee Conclusion

A wide variety of measures fall within the domain of neuroscience (e.g., direct neuroscience measures such as electroencephalography [EEG], positron emission tomography, magnetic resonance imagery [MRI], or functional MRI [fMRI] and indirect biomarkers of neural activity such as heart rate or eye blink). These measures may take multiple roles in the Army accession process including (a) monitoring test takers for constructs such as anxiety, attention, and motivation during other assessments; (b) use in research settings as criteria for evaluating other potential assessments; and (c) use as direct selection and classification assessments. Although the third role may be well in the future in terms of technically feasible and cost-effective assessment, the first two uses have near-term promise. The committee concludes that the neuroscience domain merits inclusion in a program of basic research with the long-term goal of improving the Army's enlisted accession system.

Research Recommendation

The U.S. Army Research Institute for the Behavioral and Social Sciences should pursue a program for investigating the potential for robust and objective neurophysiological biomarkers that can serve to refine and augment assessments currently in use or under development for future utilization. These biomarkers may include, among others, eye tracking, physiological reactions (galvanic skin response, cardio rhythms, etc.), medium term endocrine measures (cortisol, neurochemical markers), brain activity measures, and static and functional brain imaging. This program investigating neurophysiological biomarkers should prepare to address challenges in both what to measure and how to accomplish the measurements technically, first in the laboratory setting and eventually in field settings. The program should support research in relevant biomarker development for use in the following roles:

A. Research seeking refinement of current and future Army assessments (e.g., the Armed Services Vocational Aptitude Battery and Tailored Adaptive Personality Assessment System) through a deeper understanding of the constructs measured in selection and classification testing. In this role, biomarkers may reveal underlying neurophysiological correlates of constructs of interest (e.g., cortisol as a biomarker for anxiety). Deeper understanding of physiology has the potential to differentiate complex constructs or alternatively to reveal the relative strength of measures.

B. Independent use of biomarkers as direct selection instruments or their use in combination with traditional assessments. Research should identify biomarker correlates (e.g., consistent gaze, pupilometry, reaction time in a vigilance test) of abilities and outcomes. Test stimuli or conditions for eliciting biological responses from test takers (e.g., simulated rifle drill or other novel muscle coordination task to assess parietal-dominant brain) should be developed.

In addition to these key roles, there might be other ways biomarker development could contribute to a selection and classification program:

A. Monitoring candidates for attributes such as anxiety during assessment and offering training on mitigation strategies for applicants not selected on the basis of their test scores. Such attributes can contribute to bias in test scores, and success in controlling for the effects of these attributes can result in more valid assessment. The committee expects challenges in determining whether applicants' observed performance reflects their true ability (e.g., whether appli-

cants are experiencing normal performance stress or an interfering level of anxiety). Additionally, we expect challenges in designing a simple and effective mitigation program.
B. Basic research to apply modern neurophysiological tools to model test-taker response data (e.g., response time distributions, answer patterns that may suggest unmotivated responding or intentional distortion).

Appendixes

Appendix A

Workshop Agenda and Selection of Additional Topics Considered for Workshop Agenda

AGENDA

Workshop on New Directions in Assessing Individuals and Groups
April 3-4, 2013

Workshop Goals

1. Facilitate interdisciplinary dialogue on the current and future state-of-the-science in measurement of individual capabilities and the combination of individual capabilities to create collective capacity to perform.
2. Inform the design of a maximally effective selection and assignment system.

Wednesday, April 3

8:00 am   Workshop Check-In

9:00   Welcome from the National Research Council
         Robert Hauser, Executive Director, Division of Behavioral and Social Sciences and Education

Overview of the Board on Behavioral, Cognitive, and Sensory Sciences
    Barbara A. Wanchisen, Director, Board on Behavioral, Cognitive, and Sensory Sciences

Introductions

9:30	**Workshop Objectives and Study Overview** Jack Stuster, Anacapa Sciences, Inc., and Chair, Committee on Measuring Human Capabilities
10:00	**Sponsor's Perspective** Gerald (Jay) Goodwin, Chief, Foundational Science, U.S. Army Research Institute for the Behavioral and Social Sciences
10:45	**Break**
11:00	**Setting the Stage: The Evolving Goals of Candidate Testing and Its Role in Personnel Selection** Fred Oswald, Rice University
12:00 pm	**Keynote Address: Psychometrics for a New Generation of Assessments** Alina von Davier, Research Director, Center for Advanced Psychometrics, Educational Testing Service
12:30	**Working Lunch** Jack Stuster, Chair Topic: Discussion of ideas presented in Keynote Address

1:15	**Emerging Constructs and Theory**
	Part One: Invited Presentations
	A Psychoneurometric Approach to Individual-Differences Assessment
	    Christopher Patrick, Florida State University
	The Emerging Cognitive Constructs of Working Memory Capacity and Executive Attention
	    Michael Kane, University of North Carolina, Greensboro
	The Agentic Self: Action Control Beliefs
	    Todd Little, University of Kansas

APPENDIX A   235

                Part Two: Roundtable Discussion with Committee Members and Invited Presenters

3:30   **Break**

3:45   Ethical Implications of Future Testing Techniques and Personnel Selection Paradigms
                  Rodney Lowman, Alliant International University

               Reactions from Committee Members

4:45   Conclude Day One

Thursday, April 4

8:30 am   Day Two Workshop Check-In

9:00   **Summary of Day One and Overview of Day Two**
                  Jack Stuster, Anacapa Sciences, Inc., and Chair, Committee on Measuring Human Capabilities

9:15   **Measuring Individual Differences and Predicting Individual Performance**
               Part One: Invited Presentations
               Taxonomic Structure for Thinking About Ways to Improve the Quality of Selection Systems
                  Paul Sackett, University of Minnesota
               Rethinking Interests
                  James Rounds, University of Illinois at Urbana-Champaign
               Assessing Cognitive Skills: Case History, Diagnosis, and Treatment Plan
                  Earl Hunt, University of Washington

10:15   **Break**

10:30   **Measuring Individual Differences and Predicting Individual Performance,** Continued

               Part Two: Roundtable Discussion with Committee Members and Invited Presenters

12:00 pm	**Working Lunch** Jack Stuster, Chair Topic: Continued roundtable discussion with committee members and invited presenters

12:45	**Group Composition Processes and Performance**
	Part One: Invited Presentations
	Team Composition: Theory, Practice, and the Future
	    Scott Tannenbaum, Group for Organizational Effectiveness
	Understanding and Enabling the Collective Capabilities of Teams
	    Leslie DeChurch, Georgia Institute of Technology
	Collective Intelligence in the Performance of Human Groups
	    Anita Williams Woolley, Carnegie Mellon University
	
	Part Two: Roundtable Discussion with Committee Members and Invited Presenters

3:15	**Break**
3:30	**Cross-cutting Links and Research Gaps: Roundtable Discussion with Committee Members and All Invited Presenters**
4:00	Workshop Implications *Part One: Invited Presentation* Summary of Emerging Themes Randall Engle, Georgia Institute of Technology and Member, Committee on Measuring Human Capabilities *Part Two: Reactions from Invited Presenters and Committee Members*
4:45	**Closing Comments** Jack Stuster, Chair
5:00	**Adjourn**

SELECTION OF ADDITIONAL TOPICS CONSIDERED FOR WORKSHOP AGENDA

The following list is an unprioritized selection of the topics developed by the committee through brainstorming and deliberation processes as potential topics for inclusion during the workshop. Many of the topics were included in the final workshop agenda, while others were not for a variety of reasons. Some potentially important topics were excluded due to reasons such as time limitations of the event, availability of key presenters, compatibility with broad categories selected for emphasis at the workshop, and the committee's assessment of the likely value of discussion of particular topics over others. Some key topics not included in the workshop were included in later data gathering sessions of the committee during the study's second phase, as listed in Appendix B. This list is not all-inclusive, and it does not document all of the topics considered through two years of in-person meetings, conference calls, emails, and other information sharing that occurred between committee members, invited experts, the study sponsor, and National Research Council staff in order to arrive at the contents of this final report.

Measurement Techniques
- Unobtrusive testing methods
- Bayesian modeling
- Machine learning
- Nonparametric analyses
- Context
- Quantitative group decision making

Measurement at an Individual Level
- Constructs of cognition: knowledge, reasoning, memory, speed of processing, visualization
- Biodata
- Experience sampling
- 21st century skills
- Vocational interest measurements
- Situational judgment inventories
- Situation awareness
- Implicit biases

Modeling
- Decision theoretic advances
- Behavioral economics/game theory
- Medical decision making

Information processing models
Group modeling

Methods
Asynchronous interviewing
Automatic scoring
Communication analysis
Computational linguistics
Latent semantic analysis
Data mining
Likert scales
Clinical interviews
Sociometry
Simulations and gaming
Synthetic validation
Unproctored tests

Neuroscience and Psychophysiology
Psychoneurometrics
Blood chemistry
Biomarkers

Appendix B

Phase II Data Gathering Presentations

COMMITTEE MEETING 3

September 6, 2013

NEXT GENERATION OF TESTING: COMPUTERS, ITEM RESPONSE MODELS, AND BAYESIAN STATISTICS
Wim Van der Linden, Chief Research Scientist, CTB/McGraw-Hill

IBM'S WATSON: BACKGROUND, OVERVIEW, AND WHAT'S NEXT
Christopher Codella, IBM Distinguished Engineer, IBM

SERIOUS GAMES, SIMULATIONS, AND SIMULATION GAMES: POTENTIAL FOR USE IN CANDIDATE ASSESSMENT
Richard Landers, Old Dominion University

COMMITTEE MEETING 4

December 5, 2013

SPATIAL ABILITY
David Lubinski, Vanderbilt University

SITUATIONAL AWARENESS
Mica Endsley, Chief Scientist, U.S. Air Force

Appendix C

Biomarkers

The following text is excerpted verbatim from the 2009 National Research Council report, *Opportunities in Neuroscience for Future Army Applications* (pp. 93-95, 98) and is offered to the reader as further explanation of the linkages between biomarkers and soldier performance.

TREND 1: DISCOVERING AND VALIDATING BIOMARKERS OF NEURAL STATES LINKED TO SOLDIERS' PERFORMANCE OUTCOMES

As discussed in Chapters 3 through 6, the cognitive and behavioral performance of soldiers in many areas—training and learning, decision making, and responding to a variety of environmental stressors—has substantial neurological components. How the brain functions, even how it is functioning at a particular time, makes a difference in these and other types of performance essential to the Army's missions. The techniques used to study and understand brain functioning at all levels—from the molecular and cellular biology of the brain to observable behavior and soldier interactions with other systems—are providing an ever-increasing number of potential indicators of neural status relevant to Army tasks. The Army will need to monitor these techniques and technologies for their potential to serve as biomarkers of differences in neural state that reliably correlate with changes in performance status. To illustrate this tendency for performance biomarkers to emerge from the methods of studying the brain, three broad kinds of such methods are discussed here: genomic and proteomic markers, neuroimaging techniques, and physiological indicators of neural state or behavioral outcome.

Genetic Proteomic and Small-Molecule Markers

The development and functioning of the central and peripheral nervous systems of all animals, including humans, are regulated by genomic and proteomic factors. The genomic factors are associated with the nucleic acids of every cell. From embryonic development through senescence, the inherited genome and epigenetic[1] modifications of it regulate the expression of proteins critical for neural cell functions. This regulated gene expression produces signaling elements (transmitters), signal receivers (receptors), guidance of communication processes (axons and dendrites), and cell–cell recognition materials.

Known genetic markers may, for example, allow identification of individuals at greater risk of damage from exposure to chemical agents or more likely to succumb to post-traumatic stress disorder. The cost of genetic tests is likely to decrease substantially in the next decade, while their effectiveness will increase markedly. Of the 20,000-25,000 genes in the human genome, more than 100 are involved in axonal guidance alone (Sepp et al., 2008). At least 89 genes have been shown to be involved in the faulty formations of axon's myelin sheath (dysmyelination), associated with the development of schizophrenia (Hakak et al., 2001). Understanding the human genes associated with developments of the brain and peripheral nervous system can shed light on differential human susceptibilities to brain injury and may aid in predicting which pharmacological agents will be useful for sustaining performance. The Army should position itself to take advantage of the continuing scientific progress in this area.

A proteomic marker (a type of biomarker) is a protein (generally an enzyme) whose concentration, either systemically or in specific tissues, can serve as a reliable and readily measurable indicator of a condition or state that is difficult or even impossible to assay directly. Small variations in gene structure (polymorphisms) are often associated with differences in concentration of a particular individual, so there are important linkages between genetic factors and proteomic markers. However, specific enzyme concentrations (including tissue-specific concentration) can also be influenced (upregulated or downregulated) in response to environmental factors that vary on timescales of hours, or roughly the timescale of preparation for and conduct of an Army operation. Thus, proteomic markers can vary with

---

[1] An epigenetic modification refers to changes in gene expression from mechanisms other than alteration of the underlying DNA sequence.

recent or current conditions (environmental stressors, for example) and can also reflect the genetic traits of an individual soldier.

Proteomic markers known to signal a change in vigilance or cognitive behavior include salivary amylase, blood homovanillic acid (which correlates with dopamine metabolism), and lactic acid (a metabolic product of glucose metabolism that increases as a result of intense muscle exercise). Proteomic factors associated with fatigue resistance include microtubule-associated protein 2 and the muscarinic acetylcholine receptor. Comparison of an individual's current concentration (titer) of one of these proteomic markers with his or her baseline titer could quantify one or more neural (cognitive/behavioral) states relevant to the status of the individual's current abilities.

Neurohormones and neuropeptides—biologically active molecules much smaller than proteins or the nucleic acids of the genome—are another emerging class of markers of neurological and cognitive state and of psychophysiological response to stress. A study of candidates for the U.S. Navy Sea, Air, and Land Forces (SEALs) found that candidates with strong stress-hormone reactions to behavioral challenges like abrupt changes or interruptions are less likely to complete training successfully than those with weak reactions (Taylor et al., 2006, 2007). Another example is the work discussed in Chapter 3 on oxytocin, a neuropeptide signal, which is released when an individual experiences a sense of trust (Kosfeld et al., 2005; Zak et al., 2005). Hormonal markers are easily gathered with simple blood draws. The level in the bloodstream of a neural signaling molecule such as oxytocin has at best a very indirect relationship to its level in the brain; it may be necessary to figure out how to monitor its release in the hypothalamus. The monitoring of neurohormones and neuropeptides is likely to be a powerful means of identifying individuals who are well suited to particular tasks and may lend itself to assessing candidates for Special Operations training in particular.

Neuroimaging Techniques

Neuroimaging technologies available in the 2008-2010 time frame allow visualization of brain regions that are activated during action-guiding cognitive processes such as decision making. These activation patterns enable brain activity to be correlated with behavior. These imaging technologies and techniques include structural magnetic resonance imaging for volumetric analysis of brain regions, functional magnetic resonance imaging (fMRI) for cognitive control networks, diffusion tensor imaging for transcranial fibers, and hyperspectral electroencephalography (EEG).

Applications to Soldier Training

As an example relevant to evaluation of training, fMRI scans before and after training sessions can be compared to examine changes in the brain's response to novel training-related stimuli. Novel visual and auditory inputs activate the brain in specific regions. An analysis of event-related potentials combined with fMRI before and after novel auditory cues revealed that a particular event-related potential (a P300-like potential, which is to say a positive potential occurring approximately 300 msec after a triggering stimulus) is associated with fMRI patterns of activity in the bilateral foci of the middle part of the superior temporal gyrus (Opitz et al., 1999). Only novel sounds evoke a contrasting event-related potential (an N400-like negative potential). Individuals with a strong response of the second type also have fMRI scans showing activation in the right prefrontal cortex. These observations suggest that an indicator based on combining fMRI and event-related potential could be used to assess training to criterion. At criterion—for example, when 90 percent of the appropriate responses are exhibited in response to a cue—effective training will no longer elicit a "novel-type" brain functional response or event-related potential response (Opitz et al., 1999).

Fear is a critical response to threat that can compromise appropriate action of an individual soldier or an entire Army unit. To incorporate desensitization to fear-invoking situations in soldier training, fMRI scans could be compared before and after training to determine which environments elicit fear-correlated neural activity patterns. A prime example is the response of soldiers in Operation Desert Shield and Operation Desert Storm when sensors for chemical warfare agents indicated that the environment might contain an active agent. These fear-invoking events led to significant disorganization of military units, even when the sensor warnings were false positives.

Tracking Change in the Visual Field

The ability to track dynamic changes in objects present in a soldier's visual field is of great benefit to Army personnel. Examples include the sudden appearance of a potential threat on a Force XII Battle Command Brigade and Below display and the apparent change of terrain indicating recent placement of an improvised explosive device (IED). Jeremy Wolfe of Harvard has demonstrated that the visual system must focus on only a very limited region within the visual field to detect change (Angier, 2008). To accommodate human limitations, fMRI neurotechnology could be used to detect minor changes in the visual field and correlate them with activation events in the hippocampus (Bakker et al., 2008). Related research has shown that shifts in visual attention to objects in a field of view tend to occur either as a series of microsaccades (rapid naturally occurring eye movements) or in response to cueing signals in the field of view. Recent studies suggest that the latter is more important (Horowitz et al., 2007).

Leveraging Opportunities for Neuroimaging Techniques

EEG and EEG image processing will continue to advance, and EEG will be incorporated in multimodal imaging equipment with magnetic resonance imaging and magnetic encephalography. The high-payoff opportunity here is to leverage this work to develop a sensor array that can be used on a free-moving subject. A good initial goal for proof-of-concept would be the collection of stable trace data from a treadmill runner.

For neuroimaging with near-infrared spectroscopy (NIRS), Defense Advanced Research Projects Agency (DARPA) has been active in research and development (R&D) on NIRS sensor arrays that can be worn in situ. This is an opportunity to advance a noninvasive cerebral blood monitoring tool. Expected improvements in the next 5 years include advanced designs for multichannel data collection from cortical sources. In the 10- to 20-year time frame, one R&D opportunity is to use NIRS for more accurate imagining of the deeper brain.

Physiological Indicators of Neural-Behavioral State

Physiological indicators include individual characteristics such as age, gender, muscle power, neuroendocrine effects, neuromuscular function, vascular tone, and circadian cycling. While neural information processing is primarily a result of brain functioning and can be revealed by brain imaging, the general wellness and physiological condition of the entire human organism can affect combat capability and response to threat. This is true in large part because the brain depends on nutrient input (e.g., glucose and oxygen) via the circulatory system and on neuroendocrine function involving other organ systems. (The complex interactions between the brain and other organ systems of the body were discussed in Chapters 2 and 5.)

For Army applications, physiological indicators of neural state are important because they are often more readily accessible and measurable in the field than more direct indicators of neural state derived from neuroimaging techniques. As discussed in Chapter 2 in the section on reliable biomarkers for neurophysiological states and behavioral outcomes and in Chapter 7 in the section on field-deployable biomarkers, the idea is to find a monitorable physiological condition that correlates to a neural state with sufficient accuracy and precision to be useful as a reliable sign of that state.

Often, the laboratory studies that define the neural state and establish the correlation will begin with neuroimaging techniques (such as fMRI).

REFERENCES

Angier, N. (2008). Blind to change, even as it stares us in the face. *New York Times*, April 1, p. F2.

Bakker, A., C.B. Kirwan, M. Miller, and C.E.L. Stark. (2008). Pattern separation in the human hippocampal CA3 and dentate gyrus. *Science, 319*(5,870):1,640–1,642.

Horowitz, T.S., E.M. Fine, D.E. Fencsik, S. Yurgenson, and J.M. Wolfe. (2007). Fixational eye movements are not an index of covert attention. *Psychological Science, 18*(4):356–363.

Kosfeld, M., M. Heinrichs, P.J. Zak, U. Fischbacher, and E. Fehr. (2005). Oxytocin increases trust in humans. *Nature, 435*(7,042):673–676.

National Research Council. (2009). *Opportunities in Neuroscience for Future Army Applications*. Committee on Opportunities in Neuroscience for Future Army Applications, Board on Army Science and Technology, Division on Engineering and Physical Sciences. Washington, DC: The National Academies Press.

Opitz, B., A. Mecklinger, A.D. Friederici, and D.Y. von Cramon. (1999). The functional neuroanatomy of novelty processing: Integrating ERP and fMRI results. *Cerebral Cortex, 9*(4):379–391.

Sepp, K.J., P. Hong, S.B. Lizarraga, J.S. Liu, L.A. Mejia, C.A. Walsh, and N. Perrimon. (2008). Identification of neural outgrowth genes using genome-wide RNAi. Available: http://www.plosgenetics.org/article/info%3Adoi%2F10.1371%2Fjournal.pgen.1000111 [September 2014].

Taylor, M., A. Miller, L. Mills, E. Potterat, G. Padilla, and R. Hoffman. (2006). *Predictors of Success in Basic Underwater Demolition/SEAL (BUD/S) Training. Part I: What Do We Know and Where Do We Go from Here?* Naval Health Research Center Technical Document No. 06-37. San Diego, CA: Naval Health Research Center.

Taylor, M., G. Larson, A. Miller, L. Mills, E. Potterat, J. Reis, G. Padilla, and R. Hoffman. (2007). *Predictors of Success in Basic Underwater Demolition/SEAL (BUD/S) Training. Part II: A Mixed Quantitative and Qualitative Study*. Naval Health Research Center Technical Document No. 07-10. San Diego, CA: Naval Health Research Center.

Zak, P.J., R. Kurzban, and W.T. Matzner. (2005). Oxytocin is associated with human trustworthiness. *Hormones and Behavior, 48*(5):522–527.

Appendix D

Neural Signals and Measurement Technologies

NEURAL SIGNALS

The human nervous system has two classes of cells: neurons and glia. From the research to date, it is believed that signals within the network of neurons constitute the whole of information processing that results in behavior, while the role of glial cells is to provide physiological support to the neurons. This *neural doctrine* dominates research in direct monitoring technologies.

Neurons consist of four parts: axon, dendrites, cell body or soma, and presynaptic terminals (see Figure D-1). Electrical information is transmitted to the neuron through the dendrites, proceeds through the cell body, and leaves the cell through the axon at one or more presynaptic terminals. Neurons have one axon and from one to tens of thousands of dendrites. Details of how the action potentials (the electrical signals) travel through the cell or are transmitted across the synapses can be influenced by changes in biochemistry, which may in turn be influenced by either environmental changes or the presence of external (pharmacologic) substances.

Direct Neural Signals

In these bio-electric networks, ions of sodium, potassium, and chlorine move through the cell membranes perpendicular to the propagation of the action potential down the axon. This electrical signaling allows information to be transmitted faster than ions could flow down the axon. The propagation of information is similar to the wave traveling down a length of a rope

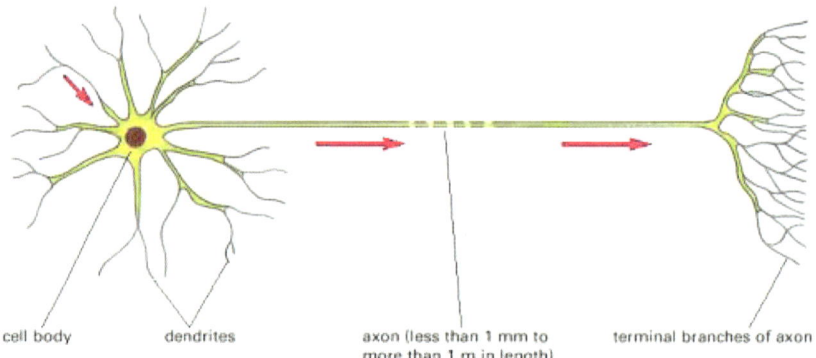

FIGURE D-1 A typical vertebrate neuron. "The arrows indicate the direction in which signals are conveyed. The single axon conducts signals away from the cell body, while the multiple dendrites receive signals from the axons of other neurons. The nerve terminals end on the dendrites or cell body of other neurons or on other cell types, such as muscle or gland cells" (Alberts et al., 2002, p. 638). The signal of principal interest for monitoring the electrical activity of a neuron is the axonal firing (travel of action potential from the body to the axon terminals).
SOURCE: Alberts, B. (2002). *Molecular Biology of the Cell.* New York: Garland Science. Reproduced by permission conveyed through Copyright Clearance Center.

when one end of the rope is moved from side to side quickly with sufficient force. Although the wave travels to the other end of the rope, any part of the rope structure has only moved (nominally) perpendicular to the direction of wave propagation. In a similar fashion, ions flow through channels across the axon's cell membrane, changing the local membrane potential and thus propagating the electrical signal down the axon.

The signal transmission down the axon of a neuron is an all-or-nothing process. When the cell body is stimulated above its threshold level, the axon transmits the same *action potential* at the same speed and in the same direction, regardless of the extent above the threshold or the duration of the stimulus.

Action potentials have durations of 1-10 msec. Input signals can result in transmission of multiple action potentials, and thus the frequency and number of neuronal firings do vary with the input. Neurons require some time to reset between firings, which nominally is the duration of the pulse for that axon. A typical maximum firing rate is between 100 Hz and 1 kHz.[1] The duration of the action potential and the speed of conduction are properties of the axon diameter and whether the axon is myelinated.

The human brain does not process information as a traditional digital computer does. Information is moved around through pathways, and at certain neurons it is allowed or not allowed to pass down that neuron based on excitory or inhibitory dendritic signals arriving before the triggering of action potentials in that neuron. Local groups of neurons can act nearly coherently, as for example in volition of motor action like a hand movement. Detecting such coherent firing at nodes around the brain is robust both noninvasively and invasively, though noninvasive techniques currently cannot resolve firing sequences of individual neurons within such groups. For example, a surface electroencephalography (EEG) signal requires the coherent firing of tens of thousands of neurons, while electrical detection of a single neuronal firing requires that a measurement probe be placed proximal to the neuron of interest, such that the probe is closer to that neuron than to any adjacent neuron. Obviously this requires opening or mechanical penetration of the skull, and that is outside the parameters of application for widespread assessment. Thus, this limitation of noninvasiveness precludes measurement (detection) of individual neuron firings.

Living neurons in an active tissue are always active at a minimal level, firing even in "resting state." Changes in the frequency of firing imply that a given neuron is currently involved in the processing of information. Bulk changes in local field-potential oscillations imply that several neurons are active. This is the baseline signal seen in the noninvasive direct measurement techniques of EEG and magnetoencephalography (MEG).

Resting state brain activity is an area of current basic research. Global patterns of activation recorded during these "baseline signal" conditions exhibit coordinated behavior when measured with functional magnetic resonance imaging (fMRI; see below and Barkhof et al., 2014, for details on fMRI). Pathologies or individual traits could eventually be indicated by modified connectivity patterns.

Purposeful brain activity leads to activation patterns different from those of the resting state. Movement as simple as an eye blink involves signal communication through a million neurons. Detecting single firings of individual neurons is a difficult process because the signals are weak to start with and are not isolated from the rest of the electrical activity within the brain. Large groups of coherent neurons, perhaps a few thousand to

---

[1] A hertz is one (firing) cycle per second, so a 100 Hz maximum firing rate, for example, would mean the neuron can fire up to 100 times per second. A firing rate of 1 kHz would mean 1,000 action potentials per second.

tens of thousands all firing at once in relation to an external event, are the most studied of single firing signals.[2]

The preceding discussion is a greatly simplified version of the electrical dynamics of neuronal firing. For instance, it does not include differences between axon and dendrite signals or the transmission of signals across a synapse. Complete discussions of underlying electrical signals in the nervous system are provided by Huettel and colleagues (2004) and Kandel and colleagues (2000).

Indirect Neuronal Signals—Energy Use

The brain activity mentioned above is a complex chain of ionic motion within the central nervous system. Ion movement within and between cells as ion channels are activated consumes energy. Replenishing the energy supply in brain cells requires the conversion of blood-borne oxyhemoglobin to deoxyhemoglobin. The rate of oxygen consumption in a localized volume varies based on local neural activity. The circulatory system compensates for changes in energy demand by increasing or decreasing both the flow rate and volume of blood, regionally and locally. Local energy demand, expressed in the capillary beds, will alter the rate at which oxygen is metabolized, called the cerebral rate of oxygen metabolism, which is abbreviated as CMRO2. When brain activity increases in a region, the circulatory response, called the hemodynamic response, will be increases in flow and volume, while the local areas increase CMRO2.

The hemodynamic response consistently provides an excess of oxygen over what is required, and this results in some oxyhemoglobin traveling through the capillary bed and local venous structure without being converted into deoxyhemoglobin. Oxyhemoglobin and deoxyhemoglobin have different magnetic susceptibilities and different infrared spectra. The hemodynamic response, by changing the net ratio of oxyhemoglobin to deoxyhemoglobin in the local venous structure, thus changes the local magnetic susceptibility and local infrared resonance spectra around focused brain activity. This complex chain reaction is called the Blood Oxygen Level Dependent (BOLD) effect (Ogawa et al., 1992). The BOLD effect leads to a method to indirectly measure local brain activity by monitoring the hemodynamic response using magnetic resonance imaging (MRI) or near-infrared spectroscopy (NIRS).

The BOLD response is a marker of the energy used locally by the coherent firing of large numbers of neurons. The BOLD response to any event peaks about 4-6 seconds after the event occurs, limiting the applications

---

[2] "Single firing signal" here means a single peak of combined electrical activity relative to an event. This is not necessarily the same as single firings of each neuron contributing to the peak.

for which monitoring these signals and their associated delay may be useful. Furthermore, person-to-person variation in distributed signals shows significant differences in regions activated (Hancock and Szalma, 2008), although there is evidence that these intersubject variations are stable over time for the same subjects (Miller et al., 2002). For use in a selection process, the brain activity signal via BOLD fMRI in an individual would need to be proven to be robust and reliable with respect to the range of environmental conditions (e.g., variations in room temperature) typically encountered during assessment.

MEASUREMENT TECHNOLOGIES

There are four noninvasive measurement technologies currently in widespread use in monitoring brain activity. The two direct-measure technologies are EEG, which detects mainly surface currents from relative voltage changes at or just below the scalp, and MEG, which detects near-scalp magnetic fields associated with neural pathway current throughout the brain, but mainly near-surface parallel and perpendicular current flow. The remaining two technologies are indirect measures that monitor the BOLD response either through rapid successions of whole brain MRI scans (using fMRI) or with NIRS.[3] The purpose of this section is to explain the capabilities and limitations of currently available technology, thereby demonstrating the feasibility of near-term possibilities to apply neuroscience in enlistment accessions as well as advances necessary for neuroscience to contribute to accurate, efficient, and mass-administrable assessments.

EEG

The primary technology used for modeling the electrical activity of the brain is also the oldest. EEG was first described in 1929 (Berger, 1929) and now exists in several derivative forms. Traditional EEG uses electrodes at the surface of the scalp to measure and amplify differences in electrical potential between points above the cortical surface and a fixed reference, such as the average reading from the ear lobes. Neuronal activity is fundamentally ionic motion in solution. Firing neurons produce the *primary current*, while induced charged-particle motion outside of the neuron is lumped together as volume currents. A noninvasive technology can only measure

---

[3] NIRS is occasionally referred to as functional NIRS or fNIRS. NIRS using multiple sources to produce three-dimensional images of internal changes in blood flow is occasionally called diffuse optical tomography. However, "diffuse optical tomography" is a more general technological term that can also refer to methods such as using visible-light laser excitation of tissue and very high resolution imaging of internal blood vessels (from inside the vessel).

the net effect of primary plus volume currents at the surface of the scalp. In EEG, orientation of the primary currents is not detectable.

Traditional EEG data are analyzed by breaking up the spectrum of combined frequencies into several bands between 0.5 and 100 Hz. A derivative form of EEG developed in the late 1930s is called evoked potentials, or EP. In EP, scalp data is averaged over several electrodes time-locked to a stimulus (Davis, 1939). Similar to EP are event-related potentials, which are measured in a similar fashion but not averaged like EP signals. Both of these methods record summed electrical activity of nominally 50,000 local neurons. Thus, large coherent group spiking activity[4] is required to produce appreciable signal.

Current EEG technologies are fast enough to capture signals of interest, making it a viable measure for research on performance as well as for use in direct selections. For example, if future technology such as phased array high impedance antennae makes localization of multiple person unobtrusive EEG recording possible, the measurements are unlikely to be any more precise than the current capabilities of an EEG via scalp electrodes. Therefore, conducting research on ASVAB/TAPAS[5] test takers using currently available EEG capabilities would indicate whether investments to develop technology for unobtrusive mass administration should be expected to yield a capability for performance assessment or direct selections.

MEG

MEG experiments rely on detection of extremely small magnetic fields produced by the time-varying neuronal currents in brain activity. Some direction information is available from MEG recordings—mainly separating primary currents flowing perpendicular to the scalp from current flowing parallel to the surface.

The signal strengths are measured in hundreds of femtotesla.[6] Typical signals are about 100 million times weaker than Earth's static magnetic field, so measurements are carried out in well-isolated chambers. Only superconducting quantum interference device magnetometers (SQUIDs) can detect such signals, and these devices and therefore the sensory apparatus requires liquid helium cooling. Hence, MEG recordings are not envisioned to be practical outside of the laboratory in the near future. However, the

---

[4] Spiking activity is the term used for recognition of action potentials. Spikes are fast and easy to recognize with electronic triggering circuits, while more complex waveforms require additional processing.

[5] The ASVAB is the Armed Services Vocational Aptitude Battery; the TAPAS is the Tailored Adaptive Personality Assessment System. Both are discussed in Chapter 1 of this report.

[6] A femtotesla (fT) is 10^{-15} tesla. The tesla (T) is the metric unit of magnetic flux density, equal to one weber of magnetic flux per meter squared.

field frequencies are defined, and theoretically, with future technology to detect femtotesla-scale fields and provide a shield from magnetic-flux noise from the environment, MEG recordings could be possible in assessment settings.

The main promise of MEG, whether in the laboratory for use in basic research or in real-world assessments, is its high temporal resolution and good spatial resolution, especially when combined with EEG information. Multimodal temporal resolution on the order of milliseconds can be combined with a spatial resolution of millimeters or even finer. Of course, detailed methods for combining EEG and MEG measurement are a major challenge; in current research, the acquisitions and analyses are done separately. Analyses are accomplished using either traditional approaches of frequency power analysis or by locking an average signal to the onset of an event cue to search for an event-related localized activity peak (similar to the event-related potentials method used for EEG-only data).

MRI and fMRI

MRI works by a simple excitation and relaxation of the spin state of protons in the nuclei of hydrogen atoms. When molecules containing hydrogen are placed in a strong static magnetic field, a small but detectable number of hydrogen protons align their intrinsic spins along the direction of this external field. An applied radio-frequency (RF) pulse near the resonant frequency of hydrogen protons, 42.6 MHz/tesla or 128 MHz at 3 tesla, knocks the spins perpendicular to the external field, and their relaxation back to ground state releases RF energy in patterns that can be reconstructed to show both the composition and distribution of any hydrogen-rich material. The resonant frequency is a direct function of the local magnetic field, defined by the Larmor relation: $\omega = \gamma B$, where ω is the frequency of precession, B is the local magnetic field, and γ is a constant of the material (42.6 MHz/tesla for bare protons, as mentioned above).

Small perturbations to the static field will change the resonant frequency. By applying a small gradient to the static field—for example, 20 millitesla/meter along the z-axis—and limiting the bandwidth of the RF excitation signal to $\delta\omega$, one may select a slice of the brain perpendicular to the z-axis for excitation to δz. A change in the gradient field will change the position of the excited slice for the next excitation. Similar gradients in the x- and y-directions can limit the excitation to a single small volume of brain tissue. In current MRI instruments, these gradient fields are produced with electromagnets and the series of time-dependent imaging gradient manipulations is called the *scan sequence*.

A free hydrogen atom (H) would produce a resonant signal slightly different from the signal from a bare proton, due to the local field changes induced by its valence electron. Hydrogen gas (H_2) would produce a still different frequency since the local field around each proton is altered by the two shared electrons. Water molecules (H_2O) contain two hydrogen atoms and an entirely different "electron shield" than either H or H_2 and thus possess another, slightly different, resonance frequency. Fats and other lipid molecules, which are important cell-structure building blocks, have long chains of hydrocarbons, and the resulting ensemble of electron screening produces a wide peak that is substantially shifted in frequency from that of water.[7]

Brain gray matter and white matter have different macroscopic lipid content and can thus be differentiated in an MRI scan. Different signals also arise in bone, cerebral-spinal fluid, and internal tissue structures of various other organs. Unlike x-ray based technologies, MRI scans can be optimized to contrast any of the many aspects of the physical signal, such as total density of protons, water content, lipid content, and even particle motion in advanced techniques involving diffusion or spin labeling. Using such scan sequences, which take several minutes, one can construct very high resolution images of gray and white matter structure for comparison with, and also mapping onto, a "standard brain" template to detect individual differences. This is important for the assessment of performance potential because different structure sizes have been linked to different behaviors and abilities. For example, larger hippocampal volume has been related to visuospatial memory capability in large-city taxi drivers (Maguire et al., 2006) and reduced medial prefrontal cortex volume has been related to schizophrenia (Mathew et al., 2014). Furthermore, models are under development to explain these differences in brain structure, but for the purposes of this study, the correlations between structure and behavior could be important for selection.

A series of fast scan sequences, typically collecting an entire brain volume at a resolution of 3 mm^3 in 2 seconds, that are calibrated to optimize detection of the BOLD signal will show the dynamics of brain functioning under the specific internal or applied conditions at the time of scan; this is known as a functional MRI, or simply fMRI.[8] The major advantages of fMRI are unmatched three-dimensional spatial resolution compared

---

[7] Frequency detection sensitivities in MRI are very good, and "substantial" here means about 3 parts per million. The frequency shifts caused by imaging gradients range in the parts per thousand.

[8] Specifically this is T2* Echo-Planar Imaging, also called BOLD EPI, Gradient Echo EPI, or BOLD fMRI. This approach is used in well over 90 percent of published functional studies, although there are more advanced techniques that concentrate on smaller portions of the hemodynamic signal. For example, Spin-Echo EPI will provide a higher localization within the gray matter but at the cost of a loss of 90 percent of the signal amplitude.

to other noninvasive imaging methods and complete skull penetration, making it the only imaging modality to unambiguously detect limbic activations important for determining emotionally laden neuropsychological states.

A long-term prospect, likely in the 20-40 year timeframe, is that combined low-field MRI and MEG technology could detect neuronal firing deep in the brain and with high temporal accuracy. Initial experiments indicate some level of feasibility, but there is substantial development work required in room-temperature, low-field magnetic field detection devices, such as atomic magnetometers, and in signal processing algorithms to sift through the substantial electromagnetic background (Kraus et al., 2008; McDermott et al., 2004). It is thought that such future devices, as well as those that might alter the atomic nuclei observed by MRI to nuclei of sodium, calcium, potassium or another element with a nonzero magnetic moment, could be operated by minimally trained technicians, the way Army medics are trained to operate medical imaging equipment for limited applications, or research assistants are trained to acquire EEG data from subjects in a sleep center. Future uses of MRI and fMRI include measurement of the Big Five personality traits and other meta-traits, which could expand upon current research to assess dual and multiple task performance with fMRI. If this research identifies brain patterns indicative of performance capability, then such tests and responses could subsequently be utilized in assessment processes.

NIRS

NIRS is an additional technology to monitor the BOLD effect noninvasively. The NIRS signal correlates with localized γ activity. (EEG measurements typically divide neuronal firing frequencies into spectra, and the relative power in five bands—0-4 Hz [Δ band], 4-8 Hz [θ band], 8-13 Hz [α band], 13-30 Hz [β band], and 30-100 Hz [γ band]—are calculated. Localized γ activity refers to an increased signal in the EEG γ band.) Studies have shown that NIRS correlates well with the fMRI signal in animal models, although with reduced coverage and lower resolution (Chen et al., 2003). This lowered resolution greatly affects the reproducibility of the technique. A recent study involving reading a preference decision from a subject in single trials only attained 80 percent accuracy (Luu and Chau, 2009). NIRS measures BOLD responses near the surface, so anything that fMRI can measure that occurs in the frontal cortex or other near-scalp regions can be detected currently using NIRS for less expense than is associated with fMRI tests.

Transcranial Doppler Sonography

Transcranial Doppler sonography measures increased blood flow through carbon dioxide–induced vasodilatation. The level of carbon dioxide, which is a byproduct of localized increased metabolism and also an indirect measure of neural activity, has been shown to correlate with the level of a subject's vigilance (Warm et al., 2008).

Ocular Measurements

Measurements involving eye fixations, dwell time (temporal length of a fixation), and pupillary changes are well-established metrics of workload in visual searching tasks (Backs and Walrath, 1992). Additional measures of ocular changes include blink rate, blink duration, blink latency, and eye movement. Moreover, fixations include small high frequency variation in the eye position that are modified by attention (Steinman et al., 1973). These are recorded using one of various types of either eye-tracking devices or electrodes to measure an electrooculogram. Eye-tracking data include the position of a fixation and the time of each eye movement (or saccade), whereas an electrooculogram only identifies the time that the muscle controlling eye blinks or eye position was activated.

Other Measures

Spontaneous eye blink rate (SBR) is correlated with dopamine activity in the brain (Blin et al., 1990; Dreisbach et al., 2005) and can therefore be used as an indirect objective measure of stress variance. SBR is an ideal biomarker for stress, as changes in dopamine activity can be indirectly tracked by a video recording of the individual's eyes. Advanced image analysis can perform facial recognition based on naturalistic video captures, and automated eye monitoring can calculate SBR (Jiang et al., 2013). Therefore, it is likely that a robust technique can be developed to determine SBR from naturalistic video recording.

The main human glucocorticoid to be monitored is cortisol. It can be measured in blood, saliva, or urine samples (McWhinney et al., 2010).

REFERENCES

Alberts, B., A. Johnson, J. Lewis, M. Raff, K. Roberts, and P. Walter. (2002). *Molecular Biology of the Cell.* (4th ed.) New York: Garland Science.

Backs, R.W., and L.C. Walrath. (1992). Eye movement and pupillary response indices of mental workload during visual search of symbolic displays. *Applied Ergonomics,* 23(4): 243–254.

Barkhof, F., S. Haller, and S.A. Rombouts. (2014). Resting-state functional MR imaging: A new window to the brain. *Radiology, 272*(1):29–49.

Berger, H. (1929). Über das elektrenkephalogramm des menschen. *Archiv fur Psychiatrie und Nervenkrankheiten, 87*(1):527–580.

Blin, O., G. Masson, J.P. Azulay, J. Fondarai, and G. Serratrice. (1990). Apomorphine-induced blinking and yawning in healthy volunteers. *British Journal of Clinical Pharmacology, 30*(5):769–773.

Chen, Y., X. Intes, D.R. Tailor, R.R. Regatte, H. Ma, V. Ntziachristos, J.S. Leigh, R. Reddy, and B. Chance. (2003). Probing rat brain oxygenation with near-infrared spectroscopy (NIRS) and magnetic resonance imaging (MRI). *Advances in Experimental Medicine and Biology, 510*:199–204.

Davis, P. (1939). Effects of acoustic stimuli on the waking human brain. *Journal of Neurophysiology, 2*:494–499.

Dreisbach, G., J. Muller, T. Goschke, A. Strobel, K. Schulze, K.P. Lesch, and B. Brocke. (2005). Dopamine and cognitive control: The influence of spontaneous eyeblink rate and dopamine gene polymorphisms on perseveration and distractibility. *Behavioral Neuroscience, 119*(2):483–490.

Hancock, P.A., and J.L. Szalma. (2008). *Performance Under Stress*. Aldershot, England; Burlington, VT: Ashgate.

Huettel, S.A., A.W. Song, and G. McCarthy. (2004). *Functional Magnetic Resonance Imaging*. Sunderland, MA: Sinauer Associates.

Jiang, X., G. Tien, D. Huang, B. Zheng, and M.S. Atkins. (2013). Capturing and evaluating blinks from video-based eyetrackers. *Behavior Research Methods, 45*(3):656–663.

Kandel, E.R., J.H. Schwartz, and T.M. Jessell. (2000). *Principles of Neural Science* (4th ed.). New York: McGraw-Hill, Health Professions Division.

Kraus, R.H., Jr., P. Volegov, A. Matlachov, and M. Espy. (2008). Toward direct neural current imaging by resonant mechanisms at ultra-low field. *Neuroimage, 39*(1):310–317.

Luu, S., and T. Chau. (2009). Decoding subjective preference from single-trial near-infrared spectroscopy signals. *Journal of Neural Engineering, 6*(1): Epub 016003.

Maguire, E.A., K. Wollett, and H.J. Spier. (2006). London taxi drivers and bus drivers: A structural MRI and neuropsychological analysis. *Hippocampus, 16*:1,091–1,101

Mathew, I., T.M. Gardin, N. Tandon, S. Eack, A.N. Francis, L.J. Seidman, B. Clementz, G.D. Pearlson, J.A. Sweeney, C.A. Tamminga, and M.S. Keshavan. (2014). Medial temporal lobe structures and hippocampal subfields in psychotic disorders: Findings from the bipolar-schizophrenia network on intermediate phenotypes (B-SNIP) study. *Journal of the American Medical Association—Psychiatry, 71*(7):769–777.

McDermott, R., S. Lee, B. ten Haken, A.H. Trabesinger, A. Pines, and J. Clarke. (2004). Microtesla MRI with a superconducting quantum interference device. *Proceedings of the National Academy of Sciences of the United States of America, 101*(21):7,857–7,861.

McWhinney, B.C., S.E. Briscoe, J.P. Ungerer, and C.J. Pretorius. (2010). Measurement of cortisol, cortisone, prednisolone, dexamethasone and 11-deoxycortisol with ultra high performance liquid chromatography-tandem mass spectrometry: Application for plasma, plasma ultrafiltrate, urine and saliva in a routine laboratory. *Journal of Chromatography B, 878*(28):2,863–2,869.

Miller, M.B., J.D. Van Horn, G.L. Wolford, T.C. Handy, M. Valsangkar-Smyth, S. Inati, S. Grafton, and M.S. Gazzaniga. (2002). Extensive individual differences in brain activations associated with episodic retrieval are reliable over time. *Journal of Cognitive Neuroscience, 14*(8):1,200–1,214.

Ogawa, S., D.W. Tank, R. Menon, J.M. Ellermann, S.G. Kim, H. Merkle, and K. Ugurbil. (1992). Intrinsic signal changes accompanying sensory stimulation: Functional brain mapping with magnetic resonance imaging. *Proceedings of the National Academy of Sciences of the United States of America, 89*(13):5,951–5,955.

Steinman, R.M., G.M. Haddad, A.A. Skavenski, and D. Wyman. (1973). Miniature eye movement. *Science, 181*(4,102):810–819.

Warm, J.S., R. Parasuraman, and G. Matthews. (2008). Vigilance requires hard mental work and is stressful. *Human Factors, 50*(3):433–441.

Appendix E

Biographical Sketches of Committee Members and Staff

Paul R. Sackett (*Chair*) is the Beverly and Richard Fink Distinguished Professor of Psychology and Liberal Arts at the University of Minnesota. His research interests revolve around various aspects of testing and assessment in workplace, military, and educational settings. His work on issues of fairness and bias in testing includes frequently cited 1994, 2001, and 2008 *American Psychologist* articles. He has long been active in the area of the assessment of honesty and integrity in the workplace. He also publishes extensively on the assessment of managerial potential and methodological issues in employee selection. He has worked with a wide variety of public and private-sector organizations on the design and evaluation of selection and training systems. He served as founding editor of the Society for Industrial and Organizational Psychology's (SIOP) journal *Industrial and Organizational Psychology: Perspectives on Science and Practice* and as editor of *Personnel Psychology*. He served as president of SIOP, cochair of the Joint Committee on the Standards for Educational and Psychological Testing, chair of the American Psychological Association's (APA) Committee on Psychological Tests and Assessments, and chair of APA's Board of Scientific Affairs. He received his Ph.D. in industrial and organizational psychology at the Ohio State University.

Georgia T. Chao is associate professor of management at the Eli Broad Graduate School of Management at Michigan State University. Her research interests include teams, organizational socialization, career development, and international human resource management. She has additional expertise in early career expectations of emerging young adults. Recently, one of her 2013 publications in *Organizational Research Methods* won the Sage Best Paper Award in 2014 and the Society for Industrial and Organizational Psychology's William A. Owens Scholarly Achievement Award in 2015. She was elected to the American Psychological Association (APA) Council, as chair of APA's Committee on International Relations in Psychology, and as secretary for SIOP. Currently, she serves on four editorial boards. She is a member of the Academy of Management, and a fellow of APA and SIOP. She has a B.S. in psychology from the University of Maryland and an M.S. and Ph.D. in industrial and organizational psychology from the Pennsylvania State University.

Cherie Chauvin *(Study Director)* is a senior program officer at the National Research Council, working on numerous studies relevant to defense, national security, and intelligence issues. She has served as the study director for projects answering the needs of the Office of the Director of National Intelligence, U.S. Department of Homeland Security, Office of Naval Research, and U.S Army Research Institute for the Behavioral and Social Sciences; and she has contributed to studies for the U.S Army's National Ground Intelligence Center, the National Institute for Occupational Safety and Health, and the Federal Aviation Administration. Previously, she was an intelligence officer with the U.S. Department of Defense's Defense Intelligence Agency (DIA), where her work included support for military operations and liaison relationships across sub-Saharan Africa and in Japan, South Korea, and Mongolia, as well as conducting worldwide intelligence collection operations (including during deployment to Afghanistan) to answer strategic and tactical military intelligence requirements. In recognition of her service, she was awarded the DIA Civilian Expeditionary Medal, the Department of the Army Commander's Award for Civilian Service, and the Office of the Director of National Intelligence National Meritorious Unit Citation. She holds a B.S. in cognitive science from the University of California at San Diego, an M.A. in international relations from The Maxwell School at Syracuse University, and an M.S. in strategic intelligence from the National Defense Intelligence College.

Ann Doucette is director of The Evaluators' Institute, director of the Midge Smith Center for Evaluation Effectiveness, and a research professor at Columbian College of Arts and Sciences at George Washington University. She has broad experience in the management, analysis, and evaluation of diverse intervention programs; the development of accountability and outcomes monitoring systems at individual and system levels; and research methodology, data collection strategies, psychometric and measurement techniques, and applied statistical analysis, including both quantitative and qualitative approaches. Her expertise includes development of performance and outcome measurement systems that target accountability, quality monitoring, and outcomes for system and individual levels of intervention/care. Her work includes a specialized emphasis on measurement, which she considers fundamentally critical for evaluation practice, and a complex adaptive systems perspective. She has developed several assessment measurement approaches using Item Response Theory to generate measures having greater precision using brief, less burdensome instrumentation, which have the potential to lead to computer-adaptive applications and real-time data usage. She has served on several technical advisory panels including the American Psychological Association's Presidential Taskforce on Outcomes Assessment and Taskforce on Pay-for-Performance; the American Medical Association's Physicians Consortium for Quality Improvement; The Joint Commission; Hospital-based Inpatient Psychiatric Services measures; National Committee for Quality Assurance attention-deficit hyperactivity disorder and substance abuse measures; and the Forum on Performance Measures for Behavioral Healthcare and Related Service Systems. She has a Ph.D. in psychology from Columbia University.

Randall W. Engle is professor of psychology at Georgia Institute of Technology. His research focuses on cognition and brain science. He is editor of *Current Directions in Psychological Science* and has been on the editorial board of numerous other journals. His interests include working memory capacity and its relationship to attention control. He is a member and fellow of the American Psychological Association and the American Psychological Society and a member of the Society of Experimental Psychologists, the Psychonomic Society, Memory Disorders Research Society, and Sigma Xi, The Scientific Research Society. He has a B.A. from West Virginia State College, an M.A. from the Ohio State University, and a Ph.D. in experimental psychology from the Ohio State University.

Richard J. Genik II is director of the Emergent Technology Research Division at the Wayne State University School of Medicine and associate professor in the College of Engineering Department of Biomedical Engineering and School of Medicine Department of Psychiatry and Behavioral Neurosciences. His areas of expertise include the use of magnetic resonance imaging (MRI) and functional MRI to gain insight into cognitive workload in naturalistic, multitasking environments. Dr. Genik has authored over 130 peer-reviewed publications and 6 book chapters, including "Functional Neuroimaging in Defense Policy," which appeared in *Bio-Inspired Innovation and National Security* in 2010. He has a Ph.D. in physics from Michigan State University and a B.S. in applied physics from Wayne State University.

Leaetta Hough is president and founder of The Dunnette Group, Ltd., in Saint Paul, Minnesota, and chief science officer of HirePayoff™. Previously, she cofounded Personnel Decisions Research Institute and served as president of the Federation of Associations in Behavioral and Brain Sciences (FABBS; 2008-2009) and president of the Society for Industrial and Organizational Psychology (SIOP) (2005-2006). She was general chair of two SIOP Leading Edge Consortiums: *Enabling Innovations in Organizations* and *Leadership at the Top*. Her expertise includes the development of staffing, training, and performance management systems; she specializes in developing measures for hard-to-measure individual-differences and outcome variables and in creating tools to evaluate a candidate's characteristics such as personality, interest, and cognitive ability essential for success in the workplace while mitigating adverse impact against protected groups. She is coeditor of the four-volume *Handbook of Industrial and Organizational Psychology* and lead author of the personality chapters in the *Comprehensive Handbook of Psychology* and the *Handbook of Industrial, Work and Organizational Psychology*, as well as lead author of the personnel selection chapter in the 2000 *Annual Review of Psychology*. Three of her articles are reprinted in *Employee Selection and Performance Management*, a book consisting of articles psychologists identified as the seminal publications in the past 100 years in the areas of employee selection and performance management. She has a Ph.D. in industrial and organizational psychology with concentrations in differential psychology, measurement, and personality from the University of Minnesota.

Patrick C. Kyllonen is senior research director of the Center for Academic and Workforce Readiness and Success at Educational Testing Service (ETS). The Center directs (a) ETS's Next Generation Higher Education Assessment and its Workforce Readiness initiatives; (b) large scale student, teacher, and school questionnaire research and development for the National Assessment for Educational Progress and the Programme for International Student Assessment; and (c) 21st century skills assessment and development research. Before joining ETS in 1999, he was technical director of the Air Force Research Laboratory's Manpower and Personnel Division. His research has focused on the measurement of human abilities, working memory, learning and skill acquisition, psychomotor abilities, personality assessment, computer-based testing, and psychometrics. More recently, he and his colleagues have been investigating affective and noncognitive mediators of educational success and job performance, along with associated new assessments and delivery modes. He has a B.A. in experimental psychology from St. John's University and a Ph.D. in educational psychology from Stanford University.

John J. McArdle is professor of psychology and gerontology at the University of Southern California. Previously he was a faculty member at the University of Virginia, where he taught quantitative methods from 1984 to 2005. He was also director of the Jefferson Psychometric Laboratory and a visiting fellow at the Institute of Human Development at University of California, Berkeley. Currently, he is director of the National Growth and Change Study, a longitudinal study of cognitive changes with age in the entire United States. His research, which has focused on age-sensitive methods for psychological and educational measurement and longitudinal data analysis, includes published work in factor analysis, growth curve analysis, and dynamic modeling of adult cognitive abilities. He has a B.A. in psychology and mathematics from Franklin and Marshall College, Pennsylvania. He has both an M.A. and Ph.D. in psychology and computer sciences from Hofstra University in New York; he received his postdoctoral training in psychometrics and multivariate analysis at the University of Denver, Colorado.

Frederick L. Oswald is professor of industrial and organizational psychology at Rice University. His expertise and published research focuses on personnel selection and workforce readiness, specifically on how to measure, model, and predict performance, turnover, and satisfaction from both individual-level and group-level characteristics (ability, motivation, interests, race/ethnicity) within various employment, military, and educational settings. He also publishes methodological research dealing with meta-analysis, measure development, and psychometrics. He is currently associate editor of the following journals: *Journal of Management, Psychological Methods, Research Synthesis Methods,* and *Journal of Research in Personality*. He also currently serves on ten editorial boards and is the research and science executive officer of the Society for Industrial and Organizational Psychology (SIOP). He is a fellow of the American Psychological Association, SIOP, and the American Psychological Society. He received his Ph.D. and M.A. in industrial-organizational psychology from the University of Minnesota and his B.A. in psychology from the University of Texas at Austin.

Stephen Stark is associate chair and an associate professor of industrial and organizational psychology at the University of South Florida. His research focuses on improving the measurement of noncognitive constructs, such as personality, in high-stake environments, computerized adaptive testing, differential item functioning, and methods for detecting aberrant responding (e.g., "faking") on high-stakes tests. He is a senior fellow of the Army Research Institute University Consortium and a fellow of the Society for Industrial and Organizational Psychology and the American Psychological

Association (divisions 5 and 14). He is currently coeditor of *International Journal of Testing* and serves on the editorial boards of *Applied Psychological Measurement, Journal of Applied Psychology,* and *Journal of Business and Psychology.* He has a B.S. in physics from the University of New Orleans and an A.M. and Ph.D. in industrial and organizational psychology with a minor in quantitative psychology from the University of Illinois at Urbana-Champaign.

William J. Strickland is president and chief executive officer (CEO) of the Human Resources Research Organization (HumRRO) in Alexandria, Virginia. Before his appointment as CEO, he spent more than 10 years as a HumRRO vice president, directing its Workforce Analysis and Training Systems Division. Before joining HumRRO, he served in the United States Air Force and retired with the rank of colonel; in his last assignment, he was the director for Air Force human resources research. He is a fellow of the American Psychological Association, past president of its Division of Military Psychology, and served for 6 years as that division's representative on the APA Council of Representatives. He currently serves as a member-at-large on the APA Board of Directors. He is a graduate of the United States Air Force Academy and earned a Ph.D. in industrial and organizational psychology from Ohio State University.

Tina Winters is an associate program officer at the National Research Council, where she has played an integral part in dozens of studies over a career spanning 20 years. She currently is a staff member for the Board on Behavioral, Cognitive, and Sensory Sciences, and she previously worked on consensus studies and other activities related to K-12 science and mathematics education, testing and assessment, education research, and social science research for public policy use. She was a coeditor of *Advancing Scientific Research in Education.*